BEING CONSCIOUS

By

Roger Taylor

PREFACE 2017

It is 28 days until my birthday when (if I make it) I will be 77 years old. I have just spent the last month in hospital as a result of kidney failure. This unanticipated vulnerability has led to my reconsidering the scope and ambition of my book. This is to be the last book I shall write, because beyond it (if I should be so lucky) there are a few other things I wish to do to complete a life. As it stands the theory of **non-mechanistic physicalism**, which is the radical treatment of the **hard problem** to which my book is committed, is, as it stands, on the page in a broad-brush treatment. The ambition at the outset was to add figurative detail to this atmospheric wash. I have now decided, if there is any merit in what I am proposing, to leave the working out of certain possible intricacies to any sympathetic others that there might be. I offer a way of approaching what is held to be difficult territory and do so in a new and provocative spirit. One of the many provocations concerns recasting our thoughts about what it is for something to be physical and so recasting the concept of science itself.

Despite restricting the scope of this enterprise I do still recommend making the imaginative journey of this recasting, the picture of consciousness in the light of what my book offers, and so seeing the world in a new way, and not befuddled by a metaphysics posing as or simulating science.

25.07.17. Brighton.

Sadly, my beloved husband lost his battle for life and died on the 23rd January 2018. His final months were dedicated to this book which he considered the completion of his life's work. Tragically, his deteriorating health prevented him from publishing but myself, his children and grandchildren have together determined to fulfill his legacy.... "Roger, Dad, Grandad - here it is!"

Marlene Taylor 2018

Dedicated to his memory

OTHER WORKS BY ROGER TAYLOR:

Art an Enemy of the People

Beyond Art

Invisible Cells and Vanishing Masses

Am Am Us

Thérèse & Tiamat

TABLE OF CONTENTS

BEING CONSCIOUS

Well I am not saying to consciousness GBWY, it is just, as my years advance, I am closing in on not being conscious, ever again, and, in Cartesian spirit, it is tempting to say, if I am not conscious ever again then I am not (not now but when I am never conscious again). This though just like the Cogito is not conclusive. We say 'she is not conscious she is asleep'. But if she is dreaming to an extent she is conscious. If asleep and not dreaming, a loud noise or a shaking will awaken her and how is this possible unless to an extent she is conscious, or is this like there being an on and off switch. What though if she is in a coma? We do not know but she may well not be capable of dreaming at all, and a loud noise will not restore her to consciousness, yet it is not that she is not. And what if from being in a coma she goes on to die? We cannot say that while in the coma she was not. 'If I am dead I am not' is the stronger candidate. But if I am not conscious, ever again, this surely is tantamount to not being. Perhaps my point is that saying 'hello' to consciousness and saying 'goodbye' to consciousness depends on consciousness: without consciousness neither is possible. And of course, as previous remarks imply, the existence of consciousness does not depend

upon such higher-order abilities as being able to say 'hello' or 'goodbye'. But for physical things of a certain kind, having consciousness is a discovery within consciousness and this allows a consciousness of the end of consciousness, and so to an inclination to understand what it is one has but will sometime lose, not in the sense of being intact but missing a limb but in the sense of a complete self-erasure, erasure of the self. But this language of **the self** is dangerous although convenient, like Allen Ginsberg talking of **the soul** but going on to say '***I mean that which differs man from thing, i.e. person -not mere mental consciousness- but feeling bodily consciousness.***' **Ginsberg, TLS, August 6, 1964.**

That which occupies space and time is physical matter, some of which, at times, is conscious.

CONSCIOUS MATTER

WHAT IS IT TO BE CONSCIOUS?

BEING CONSCIOUS / JUST BEING / BEING DEAD / NOT BEING

CONSCIOUSNESS / JUST BEING / DEATH / NOT BEING

BEING CONSCIOUS

A BOOK ABOUT CONSCIOUSNESS and CONSCIOUSNESS OF CONSCIOUSNESS

PRELIMINARY REMARKS

As I begin writing this book I am reading Ahmad Fāris al-Shidyāq. His **Leg Over Leg.** Its mode of construction, or its way of growth, I am inclined to adopt for the composition of this book. Perhaps 'composition' is too strong a word. Perhaps the concept of 'serendipitous making' would fit better, although Walpole's arrogant discarding of Thomas Chatterton is an unfortunate association. I am thinking here not so much of my admiration of al Shidyāq, and not at all of the attractions of parody, but rather of my subject matter. The life of consciousness is not systematic in anyway, or, at least, this I will argue. Systematic focus is simply a tool of consciousness. If my book follows al-Shidyāq (or for that matter to compare with more familiar or more local territory, Laurence Sterne) it will be more like **actual** minds: conscious, desiring, feeling, sensing, thinking etc., but always a fusion of them all and always moving through time and space (before, now and after, here and there). In this way it may catch its subject or resemble it. It may even, like actual minds, arrive at something not thought before.

But is consciousness a problem to be addressed? Or can the subject be taken for granted and there's an end to it? For most people it can be left unaddressed, that's for sure. I have said to several that I have before me, potentially, an immense labour, which is a work about consciousness. Never have I been asked why I am bothering with this, especially at my age. No one has confided to me that they are troubled by the topic and that my setting about the task is for them thereby of interest. Of course I know the consensus would be different confronted by a philosophic and/or scientific community (e.g. Henry Marsh, eminent neurosurgeon, on BBC's Artsnight programme 04.03.16, said '*Being our own consciousness is a greater mystery than the Big Bang and the cosmos*') rather than my usual, limited range of acquaintances. And I must say I am a little perplexed myself to find I am committing to such an undertaking. When I finished what is **currently** my penultimate book I was of the firm conviction that it would be possible to retire from the theoretical life. This firmness soon dissolved. For one thing I had forgotten my inability to leave loose ends untied. There were several threads left frayed; previous efforts which I saw a way of tidying by being able to combine them into a novella extolling the evolutionary advantages of being unnoticed. So **Invisible Cells & Vanishing Masses** was followed by **Thérèse & Tiamat**. At the same time I kept returning to the penultimate book, partly because of my interest in promoting it, but also because one of its longest parts (which in the context of the parts of that book is known as a *cell*, and the *cell* in particular known as *Escape, Block 2 Cell 6*, and probably much too long for any considerations appertaining to the aesthetic balance of the book as a whole) was, for some of the subject matter intrinsic to it, insufficiently tightly woven, its pattern not fully established or knotted off, so not long enough in fact, but how long is a piece of theory? More loose ends then.

But before I've started is there not already a contradiction? Is this not like my playing chess against my chess-playing machine, **Kasparov**, dramatizing the encounter as challenging quite legitimately for the world championship, but finding after a few moves (the computer set at one of its more basic levels) that I am already trapped and defeated? Well not quite! The mind, and as this is only the outset this is much too affirmative, like the universe itself, like any of its parts, is a simmering cauldron, subject to the forces of chaos and order. Loose ends have to be addressed but they will unravel. Revisiting the myth of Sisyphus I suppose, although it was a long time ago I read the Camus.

However, there is much more to this task than some inner compulsion for tidiness. Although if this compulsion is a disorder of the mind it should not be underestimated as I do find myself, wherever I am, constantly straightening reality, i.e. aligning cutlery, nudging mats, adjusting picture frames, etc: the attempt to impose order, something I do not really believe in. The 'much more' springs from a feeling of loathing towards a prevailing climate within science and philosophy, and more than anything it is this which has disturbed the repose I sought from a proposed retirement dedicated only to the making and attempted ordering of marks on paper, canvas and digitally sensitive surfaces (drawing, painting and tablet virtuality). What then provokes this loathing?

The matter is complicated. To begin with I am not any kind of **immaterialist**, nor do I think that understanding reality requires any abandoning of ordinary, philosophical logic. I cannot say how in introduction I am best described (this will become clearer later) but with these disavowals I am saying that what ordinarily would be taken to be positions hostile to science and philosophy I oppose too. Perhaps what I am about approximates to reinstating some of what often is pejoratively identified as **naive** or **direct realism** and attendant, again pejorative, **folk psychology** and **folk linguistics**, and as a result countering the seduction of what in science and philosophy is smart but careless thinking and insensitive understanding (the kind of thinking which leads science, paradoxically, to, in my view, a latent metaphysics).

However, by way of introduction, let me approach the matter very, **very** simply. There is a standard format adopted by film and programme-makers when attempting to educate or introduce the public about or to scientific topics. They start with some very general, theoretical patter (leaning over backwards to make learning fun and not alienating), presented by a mix of programme professionals and scientists active in the field, to be followed by what might be mistaken for hieroglyphic text but which is in fact a string of mathematical equations. It is clear that most of the audience will be at a total loss with this discourse. The text used is presented as proof, therefore as something which is true. And, of course, this will be so unless the mathematical subject is contentious. Typically in such films and programmes the equations appear anachronistically on blackboards, the chalk marks tracing a haste driven by creative frenzy. Somehow this heightens the emotive significance of proofs putting them on a par with the 'spirituality' of art, e.g. iconic paintings by Jackson Pollock or Francis Bacon! The seeming hieroglyph then undergoes a seeming deciphering or translation. There is a movement from propositions in mathematics to ordinary language propositions about reality. But how is this? Is there not something problematic about the move from mathematical meaning to meaning of ordinary language, and why should it be thought that professional scientists are proficient at this kind of translation. In fact why should it be presupposed that such translations are possible? Numbers are not hieroglyphs. So I might chalk up on a blackboard $2x + 2x = 4x$ and then quickly cut to a sequence of meta-mathematical equations (Russellian fundamentals) and claim these to be the ultimate mathematical proof of the primary equation so that $2x + 2x = 4x$ is proven universally and necessarily true. I might then add to this, reverting to ordinary discourse, that maths therefore proves that two of anything added to two of anything will give four things, whatever those things are. Prima facie this assertion, now an assertion about reality, might seem unobjectionable. Mathematics then tells us about the structure of reality, in this simple case and presumably in much more complex cases. But this is not so straightforward. That + is equivalent to 'added to' and that x, in the equation, is equivalent to 'anything' and = is equivalent to 'gives' is to fail to identify +, x and = (in the equation) as the specific mathematical instruments that they are. Thus two reds added to two reds (in paint) do not give four reds, any more than two drops of water added to two drops of water give four drops of water, anymore than two vols of gas added to two vols of gas give four vols of gas, anymore than two noises added to two noises (say cheering at a football match) give four noises. Back in the real world the whole is often more than the sum of its parts or different from the sum of its parts.

Again by way of introduction and so very simply we get black-board hieroglyphics followed by ordinary language claims like **this opens up the possibility or makes plausible the notion of parallel universes** or **the real possibility in the future of time travel**, none of these terms being technical. Professional scientists, mathematicians say things like this, and because they wave the mathematical wand they are believed, or, at least, not objected to. There seems to be no notion that what is said outside the seeming hieroglyph gives rise to a complex range of conceptual problems. All of this is bypassed by means of an invitation to the occult, trying to seduce ordinary understanding with the temptation to entertain that reality is much stranger than common-sense supposes. Much later in this text my exposition of non-mechanical physicalism connects with ‘magical properties’, but this is in no way a claim that reality is stranger than common sense allows on the contrary it is a clarification of what common sense knows. There is a delight on the scientist’s face as the absurdity of what is said is bounced into the realm of truth by the workings of proofs. But even at this early stage just stop to consider. ‘Parallel universes’ might seem to make sense but the notion of the universe (the aggregate of all matter, energy and space) is the notion of all there is physically, how then can there be a plurality of them, surely a contradiction in terms? Parallel anyway is a spatial concept so how can there be a space parallel to all of space. There is nothing unscientific or mystical about objecting in this way, it is rather an insistence on logic. The absurdity is one of translation from apriori proof to ontology. (This is not to say that there cannot be material reality beyond the observable universe.) Or consider the notion of *time travel* which has become almost an unquestioned commonplace in popular thinking and to which some science lends credence, a credibility based on equations involving space, time, mass and the mathematical complexities of relativity. There is little reflection on the logic of a conceptual structure that spans *then, now and next,* or *being before, simultaneous with or after*. What sense can be attributed to saying that something that was then but is not now is nonetheless now in that we can go to it now, so that it is then but now. Or, impossible though it is, if we were to travel to x that was then but is not now (now), then (in an entirely different sense) x then will become different from the x then that precedes now (time travel changes the 'thens' we travel to, i.e. not the same 'thens'), but if x then was preceded by w at some other then and someone from some now had travelled to that w then, then (in that entirely different sense) x then will also be different precluding the possibility of travel to x then. The whole concept is conceptually incoherent. The matter is as bad if travelling in the opposite 'direction'. Though metaphorical we do speak and understandably of 'travelling' in time towards the future, whereas, certainly as a matter of empirical fact, we do not 'travel' backwards in time. However, the notion of time-travel is not that of living through each next moment to arrive quite naturally at some future date as we might specify it starting out, i.e. knowing that now is Tuesday and that next will be Wednesday and living through Tuesday until what is then, the now which is Wednesday. What time-travel into the future requires is that what is now is what will be! We are supposed to go from what is now to what will be but for what will be to be now, but the way of getting to what will be is without living through the intervening period that makes sense of what will be being able to be the now of some future. What will be has not happened and so cannot exist now. What has not happened does not exist. The fiction that it does by-passes critical intelligence by means of a willing suspension of disbelief.

To hasty cognoscenti these objections will incur derision believing as they will that the sophistication of concepts given ordinary language names like *time-warp* etc., have not even been scratched. Perhaps these concepts are much more sophisticated than rudimentary skirmishes can allow (certainly the concept of time is contentious between science and philosophy) but the trouble is that rarely do such discussions begin. The wand is waved and everyone is acquiescent. But it is my intention to inflict much more damage than scratches in due course. The loathing then is for a world in which meritocratic power and a veneer of cleverness conceals shallow, hasty thinking, if not ignorance, and yet expects deference. There are so many factors that are a part of this. The cultural divide between science and the humanities plays its part. Typically it is said that scientists do not write well, but what truth there is in this is not a trivial truth about a formal, stylistic difference, rather it is a difference of substance, a difference in understanding how things are, a difference in being able to grasp our variable chaos of things; that our reality is mutable (although mathematics supposes it accounts for this in the abstractions of chaos theory) but on examination I will argue this is just a case in point, and as far removed from the real as similar cases like games theory and flumes theory; all subjects requiring examination. For example, assessing whether from Turing's programming equations one can derive truths about the reality of thinking and consciousness requires a sensitivity to and conceptual creativity towards the ordinary. Turing may not have possessed this, as Wittgenstein may have had to point out to him in their talk of bridges collapsing, (for later discussion). I am not prejudging how any of these particular issues are to be argued out (they need to be returned to). I am merely expressing at this stage how appalling I think it is that so much complex theory, which gets taken seriously, is generated by a lack of initial attention to or understanding of basic concepts, and how as a result the whole world may be distorted both theoretically and practically, just as medieval religion, despite the seemingly logical intricacy of its theodicy, warped the medieval mind and its social reality as a whole. The possible problems in science are compounded by the undoubted success of science in changing the world, so that its pronouncements carry undoubted, meritocratic authority. The ubiquitous dominance of artificial intelligence, information technology and robotics might lead one to suppose that the theorist pioneering these technological changes and I suppose for that matter their philosophical acolytes are in the best position to take on the so-called (in philosophy and then influentially echoed by Tom Stoppard) '**hard problem**' of consciousness. It is my intention to challenge this presumption, and to challenge it to prevent our real lives being stolen from us. What should be realised about science despite all it has contributed to the unrealised possibilities of an easier life (and in the end this is the **net** value of science) is that scientists themselves for the most part do not disassociate themselves from the existing economic order, and, instead, make positive contributions to its progress and protection. The systems of labour employed by our economic order are highly dependent on scientific theory leading to an attempted reification of social life. If the drift of science in the consciousness debate is to bring the centre of life into a reified system then a radicalised movement is required to build a realist discourse in opposition before it is too late, allowing us confidently and so without apology to live as centres of anarchic or uncontrollable freedom.

Scientists often pose as radical wizards. Their knowledge is proffered as a sort of magical enlightenment for ordinary understanding which paradoxically science characterises as gripped by something resembling witchcraft from the Dark Ages (Richard Dawkins!). The spell is the maths and a grossly insensitive version of scientific method, the wizardry is a metaphysical invasion of terrestrial commonplaces, backed by a presupposition that everything reduces to quantifiable, measurable matter. So that, for example the solidity of the real world, the world we experience dematerializes into an underlying reality of particles and charges, measurements of energy and a preponderance of empty space, seemingly not at all the world as ordinary understanding experiences it. The enticing smile of the wizard (like Brian Cox on the telly) is there before us beckoning us to, like Alice, abandon the mundane for the rabbit-hole and Wonderland, and, of course Dodgson too was a mathematician and logician. But it is a misunderstanding to suppose that the translation from one mode of discourse to another is swapping a superior for an inferior language. Reality is as both these languages describe it, and if you lack the language of solidity and materiality your grip on reality is that of a Bowie avatar, weightless and abandoned in space. And, of course, the scientist's radical posture is no more than a professional elitism and certainly not a radicalism in the traditions of real radicals like Rosa Luxemburg. Scientists are careerists, conformists, upholders of the status quo as well as the existing system of class differentiation. Science is not being used to dismantle the social order, instead it fully cooperates in building Huxley's Brave New Worlds. In their tactile dealings with reality scientists are indistinguishable from everyone else but their institutional claim to knowledge establishes a deference towards their subject and so to towards all the systems it supports. But we need to say that reality is (metaphorically) as much a slab of solid concrete as it is a worm-infested plank of wood and that one is the other and one not realer than the other. Remembering A.J. Ayer, Mozart's Violin Concerto No.3 is both scientifically measurable vibrations of catgut scratching strings and a sublime passage of the **As If** (a concept deeply embedded and elaborated in my theoretical history).

So these are some of the grounds of loathing on which this path to correction lies. And what is at issue is much more than disputed theory. The reification of consciousness is not simply a theoretical claim and if true an irresistible reality, but an assault on life itself, a social mechanisation, a pacification, a social practice and part of something much larger. The dominance of work, the dominance of time and motion is the landscape as a whole. It permeates everything. It is an ethic, it is social movement, life is for it not it for life, it drives the time and space in which we exist, nationals and migrants alike, it is the form of every slogan in political exhortation, it is the medium in which consciousness struggles not to suffocate. 'Hard-working people who do the right thing' are not thereby granted the keys of the kingdom or their children places at Eton. Instead they struggle to make ends meet, they take on debilitating debts that take lifetimes to repay; a life of anxiety and exhaustion, incarcerated in work (wage labour). This is the negative-side of work, its positive-side is accumulated capital, itself precarious. In the early stages of capitalist development the system lacked the professionalisms to dragoon its potential labouring classes, as a result this system was threatened by dissoluteness, laziness, depredation, so called, at the time, moral weaknesses (of the poor), which professional application transformed into criminality (Foucault). The reification of society had begun and on a scientific basis, driven by the reflexes of capital. But the path to correction involves dismantling the institutions of 'correction' and leads to regaining the theoretical high-ground, where lazy people are free to roam the commons and do the wrong thing, and live confident in the belief that they are the measure of all things. This is the *hard problem* of consciousness, why it matters, why it ferments loathing, and why I need to wrestle it away from those who fear our disorders of mind, which (and to partly mis-quote and entirely out of context) '*in*' our '*obscure dens, dimme caves, secret closets, merck clowdie taverns, darcke mistie victualling howses both loorckinge hydinge and absenting*' ourselves,' these unrestrained disorders ferment '*even on*' our '*ale benches, in the midst of*' our '*tippling jugges and quaffing pottes, great reasoners and talkers of devine matters & of things appending unto the same.*' (Taken from TLS 5887 review of Gerard Kilroy's **Edmund Campion**.)

My starting point with the problem will be to revisit the before-mentioned **Block 2 Cell 6** of **Invisible Cells & Vanishing Masses**. The point will be to extract from it what it contains about consciousness, putting to one side its main concern, namely the negation of determinism, although its positive account of autonomy is very much integral to the account of consciousness I want to give. The extracted content will be amplified to yield a fully articulated theory of consciousness as an **irreducible, physical property**, the nature of which does not exceed ordinary language and ordinary understanding. To this extent consciousness is not a mystery, and the efforts to make out that it is amounting to theoretical obfuscation: a means of gaining control of the conscious by denying them an existence beyond religion or science, denying them an existence within *invisible cells* (Genet, Taylor). Alongside the idea of consciousness as an irreducible physical property and fundamental to understanding this idea two other concepts, as I conceive the enterprise, will need detailing. One is the concept of **direct perception** and the other the concept of **non-representational thought**. Two concepts challenging to the axioms of science. Much more of this later.

This then will constitute the **exposition**. I add to the exposition a bundle of notes, unsystematic reflections, **commentary**, on other texts (they demonstrate the application of ideas from the **exposition** in engaging with other texts). Reading other texts has accompanied the formation of my approach to consciousness and the notes I have made have helped in clarifying positions to myself, as well as suggesting to me a range of topics that must be dealt with before the topic as a whole can be concluded. The two such texts are Cooper and Leeuwen's **Alan Turing: His Work and Impact**, and Honderich's **Actual Consciousness** (all weighty undertakings in their own right). As will become obvious there has been much reading beyond these texts but these have compelled commentary, moreover

I include these notes because in their own right they give an impression of how being conscious of something can unfold, in this case the being conscious of consciousness as a subject. My intention though is never academic, never scholarly. I have neither the time, given my age, the resources, nor the inclination to produce a definitive textbook for students or a work competing for professional advancement (those days have gone). However, it is my intention to produce a work of rigorous argument and grounded vision challenging the professional treatments of the subject of consciousness both in philosophy and science. I do this in solidarity with what I call the vanishing and unnoticed masses! Quite seriously I intend this as a substantial addition to human thought, a motivation, ambition or possibly a grand delusion always present in my own **being conscious**.

Before concluding these preliminaries there is something else. For good or ill, for enlightenment or otherwise, I have spent much of my life in philosophy. As a result I have known a number of philosophers, all, and how could they be anything else, conscious beings. Their being conscious beings a precondition of whatever identity or particularity they possessed, this precondition not being exclusive to philosophers but applying to almost all human beings. Of the philosophers known, I have known not only their works but also them as conscious beings: as a conscious being I have encountered them being conscious. However, many of those I have known, and this claim would be contested, no longer exist as conscious beings; they have died and that's an end to it. For example among those encountered who are no more are **Thomas Jessop, Alan White, Stuart Hampshire, Richard Wollheim, Bernard Williams, Gerald Cohen, Timothy Sprigge, Sidney Morgenbesser, Anthony Palmer**. In my life I think of these as lights that no longer shine: lights whose illuminations I attended to for whatever reasons, and those reasons were many. But what my exposition is about is that being dead is the death of consciousness: the death of the body is the death of consciousness. So in my account dualism has no future whatsoever. And the reasons why a dead body entails the death of consciousness is because of the physicality of consciousness. Death and consciousness are absolute limits like being and nothingness. Consciousness is physical and irreducible. There is nothing underlying which it is, nothing it can be reduced to, it is as we understand it, prima facie, it does not require translation into a more fundamental language. To appreciate what it is we have to enlarge, upgrade and so free-up notions of the physical, which for too long have been hi-jacked by professional science. The problems with science is that it confines what's physical by means of a metaphysical straightjacket. Science cannot see this because it cannot begin to countenance that it is so, and then everything gets squeezed, reduced, conceptually confined.

So, for these reasons, this work is about **being conscious** and **being dead** (the flip side of the former) but without ever approaching what might be called a philosophy of death. This work is materialist and autonomist. Understanding it will be a difficult struggle but so has been the writing of it.

EXPOSITION

A device used to further exposition in **Invisible Cells & Vanishing Masses** was to trawl through the sentences of previous cells and bring to the fore those with a bearing on the direction to follow. A sort of reminder of what went before that led, somehow or other, to the way forward: so reinforcing the direction of travel. A gathering of such sentences occurred at the beginning of **IC&VM** Block 2 Cell 6 and it is to sifting this gathering I turn first in refining the present way. Subsequently it will be necessary to sieve the remainder of B2C6 before being able to step out fully prepared. I suspect a way is only fully understood if on route we roam also its byways and hinterland. With this method it is possible to be more certain that the way is the way, and for this reason not only do I lay out the terrain from previous travel but revisit it with renewed awareness. This method also anticipates the use of commentary which will follow the exposition. The point is that there is a path and it will be followed but without resistance to temptation and its deviations there is only walking, not pilgrimage: I do envisage objective epiphany. And, to quote **David Hume** (the very same passage is quoted by **Honderich** in **Actual Consciousness**) '*We must ... proceed like those, who being in search of anything that lies concealed from them, and not finding it in the place they expected, beat about all the neighbouring fields, without any certain view or design, in hopes their good fortune will at last guide them to what they search for (1888,77-8)*'.

To begin then with something pointing to a disclaimer:- ***A post-millennial capitalism will be largely impersonal and self-generating but if all its players, winners and losers, were suddenly subtracted, negated, pow-ed, zapped, killed, it would just flap uselessly in the medium of physical geography - unless, that is, fantastically, artificial intelligence developed into real and conscious intelligence.*** **IC&VM** Block 2 Cell1.

Much of what is hostile to science could not allow this possibility. My hostility does not countenance this intransigence. I am perfectly prepared to accept that consciousness could be created in an IT lab. If for no other reason this is because the case I argue for is thoroughly materialist, although this is not say that materialism itself is in any way a finalised clarity. However, I do raise as a very serious question how it is that we are to attribute consciousness to anything. This question has to span two things. One, that any such attribution is just empty, i.e. that it attributes nothing. Two that simulation is offered as a sufficient condition because while there is nothing to attribute it is not understood what this means. To get beyond these two things would be a requirement of artificial intelligence being conscious. I will say this leads to an axiom. **Axiom 1 A theory of consciousness has to allow that consciousness could be created artificially.** (This is to say that theoretically it cannot be that emergent consciousness is confined to natural, biological process).

Next some sketchy thoughts on the emergence of consciousness:- ***...consciousness generates qualitative newness. At a certain level of organization matter becomes conscious, like a bulb or led suddenly lighting up, and consciousness, similarly to how it comes about in the first place, can transcend its inputs so as to posit what is new. All of this touches upon what is involved in ascribing freedom to conscious matter.*** **IC&VM** Block 2 Cell 3. There are a number of thoughts here all to be developed in time but central to the undertaking is spontaneous emergence. Suddenly it is there, matter is conscious. A difference between still-birth and birth. Some physical particulars are conscious. Most physical complexity is not but some is, as a matter of fact. Consciousness is a physical property, so consciousness is physical. As much a property as any other property. Without physicality there is no consciousness. Consciousness is a property of what is physical. You are out walking in a park, you think you are alone having the park to yourself when you come across a body, inert, unconscious or dead, but before your eyes it comes to life, it regains consciousness, so something more than movement but like movement in so far as movement is a property of what's physical. If what's before you had been a drone, still and then jumping, then hovering, as a matter of fact you would not think the drone was conscious, even if directed to target you, buzzing your head, like an angry wasp. The difference though is difficult to formulate yet we would have no difficulty in the standard case of distinguishing between a conscious body and a directed drone (e.g. the transitive transaction whereby the stirring body seeks to cadge a fag whereas the drone relentlessly whirs about your head) anymore than we have difficulty in distinguishing between a red thing and a blue thing (things having different properties). Is this because accompanying spontaneous emergence are the other things in the passage, **generative newness** and **freedom of conscious matter**? But consciousness is not going to be one thing. It will have many manifestations where attribution of these accompaniments will be tenuous to say the least. So is a wasp conscious or is it a drone, or like a drone, or if a wasp is conscious might a drone be? All of this needs returning to. What for now is axiomatic is, **Axiom 2 Consciousness is to be treated as a self evident, spontaneously emergent, physical property of some physical particulars.** (This is not to say that *the physical* and *physical property* are themselves to be taken for granted or are self-evident: at this stage they may only be gestures or approximations. Much more of this later though.)

Next the privacy of the mind, its possibility and what it might entail:- ... ***the mind's essential privacy. This is not the simple platitude that at any one time the mind will have thoughts, which until that time have not been expressed but later may be. The central idea is that secret thoughts are held in the mind not just as private, in so far as they are thoughts in the mind but not expressed, but, instead, they are held as thoughts not to be divulged, and that each of us has the possibility of a stock of such thoughts. If this is not so the autonomy of the mind is threatened and the concept of it lost. This threat is actual and contemporary. But this is not to argue for private languages, i.e. not to posit objects of thought which are necessarily private. It is as a matter of fact whether they are private or not. However, it is contingently possible for a mind to be privy to a body of thought never communicated, not because it can't be, but because it is held as secret. It is difficult to know how we could conceive of the thinking mind if this was not a possibility. Our whole concept of mind would otherwise have to be recast ...*** **IC&VM** Block 2 Cell 4.

A machine holds encrypted data, access prohibited without appropriate code. This data is secret data, hidden data, even classified data. This data will be held as binary code and decoded will present as data. Is this a description of a mind holding a thought as secret? 'Mind' and 'thought' here are not necessarily self-evident, it might be said they lack precision, that they lack referents. But they are part of the ordinary language we use and do not offend ordinary understanding. And in this language we will talk not just of **thoughts** that are secret or hidden but of secret **thoughts** we contemplate, we attend to, **thoughts** we hold before **the mind**, indulge in, for pleasure, pain or innumerable other reasons of fixation. This is not some binary, neural existence, stored and ready for use when triggered, dormant then presented but presented to some other. We live the secrets of the mind and approach them with desire. They are part of what it is to say we are conscious. Things that are with us but we refuse to divulge. What is axiomatic now is **Axiom 3 Consciousness holds within it a living privacy**. (This is to say consciousness that is self-conscious. Perhaps it is so that consciousness is on many levels, and consciousness can be ascribed to many beings lacking self-consciousness. Probably it is the case that being conscious implies being a being, and so if a machine was conscious it would be a being. Current machines have names but are not thought of as beings. This may be because they are modelled on intelligence (*artificial intelligence)* and despite the esteem we grant intelligence, for the most part intelligence may just be a machine-like function which minds perform. This would explain how machines which are more or less stupid in human terms, or even in biological terms, can out-perform grandmasters at chess. *Intelligence* like many of the terms employed here is only an approximate pointer: this will be for later.)

To highlight the variety that the existence of consciousness contains:- ... ***although we talk freely about animals thinking, there is probably unease in ascribing mind to individual animals unless we see some evidence of something approximating to forms of cogitation, some gap between thought and action. So very approximately we might be treading firm ground in talking about what goes on in a gorilla's mind but to wonder what goes on in the mind of an individual sparrow is for us more like fanciful fiction.*** **But it's all a matter of degree. It's all the same sort of substance but with different evolutionary profiles. IC&VM** Block 2 Cell 4. How committed should we be to the concept of mind in trying to determine consciousness? The language of mind is suggestive of some non-physical place or box, closely aligned to some concept of self. A something with an identity. Some notion of a person. Our natural talk is of things in the mind, on the mind, at the front of the mind, at the back of the mind, what shoots into mind etc. Spoken of as a box but not to be found in the head or brain. And if all of this is something other than what is physical it is not hard to share **Ryle**'s doubts in *The Concept of Mind*. But perhaps pressing the language of mind is being too exacting, is not allowing it to be just part of the ordinary language, something that helps to point to what we recognise. And perhaps here we do not need to stray too far from the language of consciousness itself. For some consciousness there is consciousness of consciousness, in this way the moments of consciousness can be linked. Perhaps we should say this linking is the mind, and that for this reason there is reluctance to (although bearing *in mind* **Axiom 3**) ascribe mind to a sparrow because any such linking goes way beyond our experience of its consciousness, whereas, empirically, there is no similar difficulty, in the standard case, in ascribing mind to humans. This consciousness of consciousness pulls in other attendant concepts, like concepts of self and identity where consciousness of consciousness makes possible the linking of conscious moments or episodes and so makes possible a history of consciousness for a numerically separate physical particular (i.e. one that has consciousness as one of its physical properties). Not that mind and self-consciousness are equivalences. Just reflecting on the concept of **a troubled mind** should convince of that, but self-consciousness might be a prerequisite for the ascription of mind. So to **Axiom 4 Mind is a superstructure generated by and resting on a base of self-consciousness**. (If this is so then mind is a condition of **Axiom 3**. And *superstructure*_and *base* are just metaphors for something requiring more concrete determination.)

And on to what science does and does not do:- ... ***the scientific programme is not only remote from the possibility of translating precisely generalized brain states into mental states, but even at the level of physical theory about physical substance, where there should be no doubt about the achievements and objectivity of the physical sciences, the determination of particulars in natural history is not really attempted by science because it is not this kind of account that scientific theory yields. The scientific possibility of staying abreast of the mind's secret life on the basis of mapped traces of electrons has to be compared with the possibility of the theory of atomic particles being able to plot, say on a beach, the movement of a particular grain of sand, or a particular***

pebble even, over changes in time -a decade, four seasons, even a day-. Scientific understanding eliminates variables; the minutia of reality is a complex of variables. This is not to say that science does not understand reality, or does not have a theory of variables: that would be an absurdity ... **IC&VM** Block 2 Cell 4.

Plot here means predict. Science would have no difficulty precisely tracking a grain of sand over time. From such observation it would be able to confirm that given conditions n ...∾, and they were given, then grain 1 of sand in space 1 at time 1 **had** to be in space 2 at time 2. What science does not do and is unable to do with scientific certainty is to predict for any particular grain of sand that there would be conditions n ...∾ and that g1 would be in s2 at t2, although, as we could all do, if conditions remained exactly the same so that the issue of variables could be set to one side, science could predict accurately that g1 would be in s1 at t2, but really this is a deduction. This is not a limitation of science as there is no expectation that science should provide biographical accounts of particulars, and this is not to say science does not forecast. The point is that its forecasts are general. Weather forecasts are not pointless and can be very accurate in the absence of complex variation. However, we do not expect a detailed continuous forecast of all aspects of weather for a particularity, for, say, a plot of land, perhaps 40 square metres or even a square kilometre. This is not to say that science does not have a means for dealing with the particular. In laboratory circumstances science fixes the variables, and so it fixes a particular situation which becomes generalisable because vagaries are fixed and so eliminated. Its generalisations will apply to the particular though, so a grain of sand or a plot of land will be subject to laws about the drift of sand and the growth of vegetation in a climate, but the precise detail of these particulars eludes and is not part of scientific forecast. The point is that we can generalise about particulars but particulars remain particular. Variation is what makes evolutionary theory retrospective rather than predictive. Of course the problems become even more complex when the particular has what a grain of sand does not have, that is, consciousness. Consciousness is the variable of variables and in those consciousnesses having the consciousness of consciousness, such conscious particulars, as argued at length in **IC&VM**, can always choose other than they choose, so removing a necessary condition of scientific forecast, i.e. determinism of effect. From this we get **Axiom 5 Consciousness is a physical property of neurobiology but neurobiological facts are not a translation of conscious life.** (It will follow from this that the private life of a particular consciousness can never be inferred from neural facts without the collusion of consciousness. What will require deep probing is the concept of physical property, and it may well be a better concept can be suggested, but whatever is suggested must emphasise physicality and this in turn requires its own illumination.)

More on privacy:- ***The essential privacy of the mind is under attack socially. The development of the human mind was a slow, evolutionary process and what anthropology classifies as early human would not have had available to it in any developed form the essential privacy of the mind. An example of a body of thought 'minded' for private contemplation would be a narrative, a story, not shaped as in literature, more an anti-narrative, a post-modernist story, joined and disjointed, direct and allusive, associated and disassociated, literal and metaphorical. At some point in historical development consciousness contained this possibility as ordinary: settling the world but with other worlds in our heads and other worlds retained for the self. When this happens it adds dimensions to any sense of self and identity. Just as the construction of stones in a landscape -the Pyramids, Stonehenge- develops social identity so constructions in the mind, specific to particular minds, develop a sense of personal identity. In the history of philosophy, rationalists and empiricists have treated the self either as an unknowable but necessary mental substratum or as some totality of ideas, but that one is oneself becomes blindingly obvi-***

ous when one has the possibility of constructing something like a narrative withheld from others. What is the self? It is something that has to contemplate this as something unique to itself. It is something that can withhold things from others, and the concept of personal freedom is integral to the possibility of this withdrawal, because it is part of what is meant by it. A private body of ideas, privately constructed and privately accessed is the birth of the self ... the concepts imply each other. And on what basis can this be denied? By denying that there is any such essential privacy that we can make and search? But for some of us at least, this denial is a limit too far, a cogito moment: a clandestine body of thought that refutes the denial but which will not be revealed for this purpose. **IC&VM** Block 2 Cell 4.

So what starts to emerge is a radical yet common sense understanding of mind and self. We can withhold, positively withhold. This is a sense of what's mine and not yours, not the others. This a lived distinction. This is not a dead encrypted barrier, just prohibition. Withholding is what we do, and so live rather than just be, it is a sense of what is encompassed and will not be divulged. It is, we say, the privacy of the mind. Not something appearing on some internal screen that we 'see', it is seeing itself, something alive, continuous through the moments. This is identity. This then is a clear meaning of self, the living option of divulging or not divulging. So the self has to identify something that may or may not be divulged. The self is this living whole. The sense of a something and the living withholding or not of this something. And this life is part of the physical order. Before us in each and every living person. The self is not then some spiritual, non-physical, Platonic entity, nor just a collection of ideas (itself an absurd way of thinking of mind, events without continuous connection) instead the self is a physical presence that is alive, a property of a physical particular, and containing indubitable privacy as its primary dimension: this its mode of living. For a machine to be conscious it would have to have an inner life of its own, lived gratuitously and which it *freely* decides to divulge or withhold, where all these terms have their full meanings in the sense that science often denies that they can (as though a conceptual absurdity). Here is the crux of the problem. So **Axiom 6 A self-conscious consciousness continually confronts the choice of divulging or withholding.** And think of autism.

But privacy of consciousness is not simply a fact, it is a contesting fact, it is a force, a movement in the world, one that threatens the world and is threatened by the world:- ***Constructions of the mind are anathema to capitalism. They can take place without the mediation of the consumer, without the activities of buying and selling. Capitalism encourages surrogates for these activities. So there are endless consumer games, and endless computer programmes where construction is simply choosing options, or sometimes just 'clicking', from inexhaustible given inputs - mimicking the naivety of the empiricist-creation theory-. And the celebrity model of human behaviour predominates, where nothing is left hidden, where everything is revealed, where everyone is encouraged to reveal themselves in a public, commercial space: space timed and paid for. With the commodity nothing is private and lives are to be lived as commodities. In these circumstances the imperatives of authenticity and expressionism serve the market. The inner life, the privacy of the mind and associated things come under attack. The individual becomes image, personality, character, persona, all recognizable and exchangeable properties and all shared with all commodities, but not the same as a concept of self, not the same as something that might be given or not given. The TV show*** **Big Brother** ***is a post-modernist spoof on Orwell's nightmares, but its participants, members of so-called free and democratic societies, are eager for the life of celebrity, eager to reveal everything, eager to hand over their lives to*** **Big Brother.** ***If for a moment any of them try to construct Winston Smith's shallow alcove they are named and shamed for not playing the game, ultimately chastised for not 'honouring' the contract. In these differ-***

ent ways the mind's essential privacy is under attack. But it is probable that it developed as an alternative to the consciousness of systems, not teleologically but accidentally. Perhaps the genetic material that produced the defence of group-systems, being instantiated in some mirroring form in the members of the group, mutated individuals into microcosms of the macrocosm, like Leibnizian monads. Empirically it is true that the sense of self is entrenched and the reification of commercial experience cannot resist interiorisation and construction which probably lead to the replication of 'Winston's alcoves' throughout society, while media led society, in its ignorance, thinks all human experience reduces to the format of **Big Brother.** *Most of us are more than the system predicts and profiles, and we know this for ourselves. It is also this that sets off alarm bells in commercial process: 'out there' so many moving around carrying unpredictable worlds in their heads.* **IC&VM** Block 2 Cell 4.

This passage suggests an enlargement of consciousness. The consciousness of systems, the consciousness of groups (i.e. not being conscious of a system or group but system/group consciousness). Strangely in ordinary consciousness it is the consciousness of individuals that first springs to mind. It is almost as if we have to turn to philosophy before encountering the enlargement: philosophy of Hegelian inclinations. There are ordinary language concepts implying recognition of the subject, so we talk of class consciousness, the spirit of a nation, the will of a people. Is a colony of bees conscious when it swarms to confront a threat? A pack of wolves conscious as they all contribute to the capture of prey? We do not have to press the concept too hard. We do use it in this way. Perhaps just an enlargement of meaning. However, the concept of a consciousness constantly suspended between the hidden and the revealed does not fit, seems out of place in a pride of lions, a pack of wolves, a swarm of bees or, in particular in this context, a tribe of prehistoric humans. So what is envisaged in this passage is an evolving self-consciousness. A movement from group consciousness to individual consciousness. Whether this is so or not and how it might have come about is here only speculative. Perhaps it is at the point of there being misfits and mystics, holy men taking to the wilderness, or perhaps at the point of written language or oral poetry that self consciousness evolves. I think it must be allowed that human thought can be active, shared, determining social behaviour without there being a socially realised possibility of stepping back from thought and action and so thinking of self thinking. So I am suggesting where we are now, where consciousness is now, is not some universal but must be some evolution we have passed through and so not ahistorical but developing and as a consequence something that might be changed, threatened! And this passage suggests, is threatened. What is threatened is not the mere fact of being conscious but the breadth of consciousness. Consumerism, the way of the world, is content with being conscious of what to buy and buying: for consumerism the limits of thought and action. Any more than this impedes the way of the world and it is possible to restrict life in this way. A manipulation of our being conscious. But regressive consciousness is an alternative world, stripped of exchange value and only potentially regained when content is revealed. **Axiom 7 Consciousness is broad, evolving, stretching and always fully engaged in a struggle with reification, where, in the first place, it finds itself.** A matter of bringing the physical to life, so matter brought to life, but confined by consumerism and fascism and other attempts to make machines out of conscious beings: an appropriate contrast in the context of this book.

Now to sieve the remainder of **IC&VM** that relates to this enquiry, namely Block 2 Cell 6, starting with material taken from the first part of the main argument of the cell, the section entitled **Determinism, the Apriori Abstraction**:- *But perhaps it is what we share with madness that makes us open to something like* **'I am self-conscious therefore I can'.** *This is to say that unreason and reason, in the Foucault senses, make the whole that we*

are, and it is this which is a clue to the meaning of self-consciousness, and it is because we are self-conscious that we can always do differently no matter how overwhelming the reasons for not doing so. Self-consciousness makes possible unreason and so the profundity of madness. Someone who always acts against their self-acknowledged best reasons for doing things is someone, who we would, in ordinary, non-clinical language, say is mad, but unreason is a choice for those who are self-conscious (necessarily?). We might say the mad affirm their ability to do differently, and reason holds them mad to do so, but this madness is part of how we all are, and so in denying that we can do differently we pretend that what is beyond the black stump -the marker post between civilisation and its antithesis- is not a continuum, not parts of a whole which is ourselves. What it is that we may now pretend may not always have been so. Foucault makes the point about C17th and C18th Europe.
'But in the seventeenth and eighteenth centuries, the animality that lends its face to madness in no way stipulates a determinist nature for its phenomena. On the contrary, it locates madness in an area of unforeseeable freedom where frenzy is unchained; if determinism can have any effect on it, it is in the form of constraint, punishment or discipline.' **Foucault** ***Madness and Civilisation.***

And '*secret self-consciousness*' is at the root of that which is alive, the essence of which is so admirably expressed by **Thomas De Quincey**, that '***inner world, the world of secret self-consciousness in which each of us lives a second life***' (De Quincey's second life being that of the Ancient Mariner). A second life and unreason going together with consciousness and freedom and being alive. Science does not replace these things by inventing simulations. Only when science has produced a machine that is **mad** will it have pulled off the trick. Only when the machine can do the opposite of what it is programmed to do, when it is capable of unreason and when it can live a second life, will the machine be conscious and alive. And this is not to say that this is impossible as the dualist would argue. Life is physical, whereas spiritual life is metaphysical and so a conceptual confusion. Living consciousness is a possible physical, material property of a material universe. Its human manifestation is accidental but it could be intentionally created, by physical beings possessing physical consciousness. So **Axiom 8 Second lives and madness are the unforeseeable freedom of self-consciousness**.

Some more from **IC&VM**, Block 2 Cell 6, **Determinism, the Apriori Abstraction**, and bearing on determinism:- ***Apriori determinism cannot be compatible with any state of affairs. It has to deny something. It has to close off some concept of autonomy. If we always have available to us an infinitude of options, though relative, and if we can always choose differently, because we can always stand back from where we are, making how we are an object of consciousness, including stepping back, then this is not compatible with an analysis of choice requiring it to be both effect and cause, although once something happens it happens one way. The truth of the proposition that once something happens it happens one way cannot be the proof of determinism.***

Some concept of freedom then is integral to understanding the spectrum of consciousness. This is not the place to press for some clarification of freedom. Minimally it is the negation of determinism. So if determinism is understood then the freedom of consciousness (here meaning those consciousnesses that are self-conscious) refuses what determinism asserts. And it achieves this because the freedom of consciousness means that conscious beings can always choose differently. Such consciousness is always being able to stand back from what it is conscious of, and because of this it always being possible to do differently, even though contrary to everything desired. Most science would deny this. To allow it, is to allow freedom. From the point of view of empirical generalisation science is for the most part correct but this does not foreclose an apriori argument. Taking the most obvious case why should we not act in accordance with our history of desire? If we are free this is compatible with our being free, but, of course, if we could only act in accordance with our history of desire we would not be free. But we all can refute this, this determinism. How can it not be possible for you to attempt the opposite of what you are doing? 'But this is perverse!' However, this is the point. For a machine to be conscious we have to be able to attribute to it potential perversity. And perversity draws to itself a cluster of concepts that would need to apply to a conscious machine, such as being amused, feeling guilty or ashamed, and being secretive. How do you make a machine that is amused? Certainly not by giving it a mechanism that makes giggling noises, or tells an ironic joke. Is there a philosophical or scientific proof of what is asserted here? Philosophically the proof is in the proper understanding of the concept of consciousness. Scientifically, i.e. empirically, evidence shows that conscious beings are unpredictable, although for the most part not. Frequently we are confronted with stories of quiet, well-behaved, conservative individuals who perform the unthinkable, the unpredictable, the unprecedented (suicide bombers, pilots who deliberately fly airliners into mountains, etc.). Freedom is terrifying, it is a 'frenzy' that is 'unchained', and, of course, this is not something that repressed society can admit. For an authoritarian society the apparatus of the state, which in our day and age includes science, will attempt to pull everything back into a determinist order. So the suicide bomber has to be groomed, radicalised, manipulated and under no circumstances can just freely decide to challenge the whole culture in which he or she has been conditioned. But the proof is in ourselves. That we can think the opposite of ourselves (fundamentally because we can make ourselves the object of our consciousness) then we can attempt the opposite of ourselves: we can do this just to prove that we can. This is our freedom, and that it is our freedom **means** we do **not** have to exercise it. I am conscious therefore I can choose otherwise. A *cogito ergo sum* moment, automatically translated on Google, and strangely, as *Conscius sum ego potest eligere aliter.* **Axiom 9** therefore **Free conscious beings can choose other-**

wise (than what they have chosen; perhaps quantum beings).

Next more from **IC&VM,** Block 2 Cell 6:- ***Models for life stare us in the face in biological evolution: the struggle for survival. Babies can be easily exterminated, easily have pins stuck into them, even so it is more difficult than pulling a plug on a computer. Babies make a fuss. They feel and express pain. A quite different order of reality than designing a computer that screeches when shut down. To make consciousness one might try to make pain. Artificial life that felt pain would confront us with a whole new horizon. To make pain we would have to overcome all the confusions of simulated and real pain.... One would not make real pain by appealing to an epistemological smokescreen of uncertain criteria and the identity of indiscernible simulation.... It is just that the making of pain points to an order of thinking quite different from the making of hypertext. Most likely the possibility of progress towards the dimensions discussed here will depend on advances in biocomputing, but the claims for DNA origami will have to do better than jokes about having created the most concentrated happiness ever on the basis of folding DNA resembling 'smiley' faces.***

If it is supposed that pain is no more than messages transmitted through nerves and pain-expressing, pain avoiding behavioural conventions (reducing to physical movements, e.g. screaming, clutching the head, reaching for the pain-killers) as it has been supposed, then it might not seem difficult to make a machine about which we might readily say that **it feels pain**, where 'feeling pain' means no more than what has been *supposed.* In some ways addressing the problems inherent here might be the crux of all the problems. Pain, our pain, the pain of others shouts at us but much scientific/philosophical discourse does not hear this. The deafness is a fear that there might be something other than what is physical. But the difficulty lies in the concept of what is physical: a concept, and this I have to argue, which is conceived too narrowly. For the most part the problem is one of supposing that the reality of the physical is confined to what can be subjected to rigorous measurement and quantification. Although science does *play* with the idea of measuring pain, whereby patients complaining of pain are asked by medical professionals to grade their pain on a scale, say 1-10, it is not thought that any real, scientific accuracy can be attributed to the scores. At best the patient's testimony is taken as a subjective and questionable indication of what treatment to prescribe, although prescribing medication that reduces the *subjective* assessment is seen as a positive behavioural outcome in its own right. Pain on the other hand reduced to detectable changes in neural transmission rates and brain activity, which drugs alter, would be regarded as the objective reality of pain. But pain without consciousness is not pain. This is to say that even if there is a high level of neural transmission and brain activity of the type correlated with pain but a conscious patient does not report pain, or subjectively does not feel pain, then the scientifically located pain is painless and therefore is not pain. Perhaps this is saying that there is no pain without *painful sensation*. It may be that science would want to link neural activity to verbal behaviour and mental activity, probably causally, and so concede the link, but would only do so where the reality of pain would be no more than these causally linked events, otherwise pain would be assessed non-existent. But, and so importantly, pain screams at us and does so physically and so empirically. Pain is physically located: in the arm, at the centre of the chest, in the tooth, from the pulled hamstring, the infected sinuses, the cancerous bowel. Pain, as we talk of it and experience it, is not something belonging to what is called *spirit*. It is not something non-physical. It is obdurately physical and yet unlike fingers and thumbs it does not survive death although in some instances it is felt to survive the removal of the location where the pain is felt (how accurately these instances are described requires scrutiny). The way we describe our pain engages with the language of the physical. Pains are gnawing, throbbing, sharp, dull, mild, acute, tight, excruciating, burning, bursting, pressing. So if it makes sense to say that consciousness is a property of the physical, then somehow pain is interwoven into this property. Consciousness and pain confront us as directly as anything in our existence: confront us as physically as anything in our existence. They are as undeniable as anything in our existence. And their confrontation of us presupposes a conscious physicality which is confronted. Perhaps what this points to is that the consciousness, which this enterprise sets out to locate or describe, has to be approached as something multi-layered, as a transparent medium, constantly self-referential in its depth. Pain is not a special case. Consciousness, as physical, is not something separate from our physicality. Consciousness is all the time a consciousness of holistic physicality, of which episodes of pain are an integral but small part of physical being. There is nothing metaphysical here and there is nothing non-empirical here. We all know all of this unless we allow metaphysical detours (spiritual, scientific) to un-rail us. So, for now, making a machine that feels pain must involve making it genuinely conscious, which means giving it depth, *the deep* (fathoms of transparency). This is to say that consciousness is not like a two-dimensional vantage point onto a three-dimensional reality or a digital reality (screens, objects and the programming of an object), but has to contain a physical awareness of physical awareness (all of which is aware of a physicality that is not aware) and the two transitively aware. Not that this anywhere near defines *the deep*. There is so much to this *deep*, so much to this transparent medium. Much more than physical sensation, much still to be ascertained and all of it together, transitively together, at all times. This a holistic account of consciousness which ridicules a tripartite division into perception, cognition and affectivity, of which the history of philosophy is the perpetrator and culprit. So, **Axiom 10 The making of pain in a machine requires a multi-layered, transparent, self-reflexive and transitive depth of genuine awareness.** Much of what is included in this axiom is metaphorical and in need of thorough exposition. Hopefully more of this will be provided in due course. For now I trust enough has been said to suggest a prima facie plausibility for the idea that consciousness and all it is, is a physical reality in its own right and has to be approached and understood as such, not reduced to something else, like a simulation of itself. **Consciousness is a fact of physical existence.** Perhaps this is **Axiom 11**. If it is a fact it is an accidental one. So, **Axiom 11 Consciousness is an accidental fact of physical existence.** (Later on in this book 'accidental' will morph into **magical**.)

As a sort of postscript to the above and reflecting on the sociology of medicalisation which highlights a conflict between theories of biomedicine and more integrationist approaches (see **Gillian Bendelow, Health Emotion and the Body** Cambridge Polity Press) one might say that a physically existent consciousness, in whose depths lurk sensations of pain, points to a way of recasting the difficulties. Both biomedical and integrationist approaches leave themselves open to the spectre of dualism: bio-medicine fixated on the mindless body and integrationists arguing for something other than the physical, namely mind and emotion, and how they play a role in determining what is physical (e.g. the experience of pain). But if consciousness and pain are themselves physical properties there is no theoretical difficulty in transitive determinations between them and what is conventionally physical. Objectivity now demands a holistic approach, and anything less would be a distortion, a truncated physicalism.

And more from **IC&VM**, Block 2 Cell 6, continuing with the theme of freedom and madness and leading into the *emptiness* of consciousness:- ***Foucault quotes*** *(Madness and Civilisation)* ***Mathey, a Genevan physician and someone who, Foucault says, was very close to Rousseau's influence.*** **'Do not glory in your state, if you are wise and civilised men; an instant suffices to disturb and annihilate that supposed wisdom of which you are so proud; an unexpected event, a sharp and sudden emotion of the soul will abruptly change the most reasonable and intelligent man into a raving idiot.'** ***Everything neat, ordered, mechanistic,*** *worked out to the last mathematical detail, human beings at last understood, and then you are John Nash. And, Rousseau was such a case, but the point is that we are all such cases. In all our lives there are these moments of madness, not because some cause overwhelms us, but because we can choose anything no matter how boxed in. This is exceedingly dangerous for society. It is better, it might seem, to hold that the mad are possessed, an infrequent few, to be confined and sedated, rather than admit that this is our condition, the human condition. That* **autonomism** *is true means we can strike out in any direction, and this is the refutation of apriori determinism, and also, if true, the refutation of aposteriori determinism. That we can strike out in any direction requires moralists to have very strong arguments to win our agreement and for us to limit our acts. In fact none of the arguments are strong enough ... Instead moral control sets up moral institutions, which make use not of argument but of fear and coercion to achieve restraint and confinement. These fear induced incarcerations of the mind are themselves extremely powerful despite our freedom, and so the descendents of African slaves in Western Society gave thanks to a Christian god for the abolition of slavery. Autonomous agents are amenable to control by fear, but the proof of autonomy is that madness is an escape from real fear; displacement for a time into other worlds. Apriori determinism cannot allow that 'choice' is an indubitable concept of freedom; instead it requires that choice be a cause and therefore an effect -as it would not be a first cause-. The difference between these alternative conceptions of life cannot be settled apriori. If the sense of the indubitable is delusory then confirmation of this will need to be proved empirically and so cannot be settled apriori.*

The truth of autonomism, according to this argument, has to be empirically grounded. **IC&VM** contains arguments which dismantle the pretences of apriori determinism. If determinism is true it has to be established empirically, and because of this, after dealing with apriori determinism, **IC&VM** examines the scientific case for determinism (more of this later), finding, in the process, that the scientific case is often nothing of the sort but is constantly falling back on apriori determinism. But before the detailed argumentation, which is necessary, this passage from **IV&VM** points to what it supposes is empirically obvious to all of us who are conscious beings. For example, we all know from experience that we can bring to mind some specific identity, say Pol Pot, more or less at any time or place. If there was some scientific argument opposing this we would do well to be dismissive of it. Of course I might be hard pressed with other things, things of major importance, so that bringing Pol Pot to mind would be inconvenient or something that would not *cross the mind.* But on what grounds could we refuse this possibility about ourselves? And this would be a case of choosing differently. So we might say nothing stops us from thinking anything, and not '*if we like*', but '*whether we like it or not*'. Of course we can sense the strained counter arguments muscling to the fore. Like, and this would only be minimally relevant, '*what can be thought of will be determined by inputs from experience*' but this I will not go over again as it has been extensively refuted in dealing with the empiricist-creation theory both in my **Beyond Art** and **IC&VM**. What can be sensed here is that any such argument is already posturing towards closing doors apriori. It is empirically indubitable we can think of anything at all, more or less, even if our minds have been invaded by coercive, moral censors. This is a simple example of our being able always to choose differently, in fact an explanation of our being able to choose differently. Normally we do not do this, and why should we? To do this, to choose differently and being able to choose anything whatsoever leads to the vocabulary of madness. This language points to the validity of autonomism. Science (and I'm inclined to say *of course*) attempts to mechanise madness. So the John Nash case is a chemical imbalance of the brain and can be altered by chemical intervention, and similarly the suicidal departures of **Primo Levi** and **Nicholas de Staël.** But this dehumanises the subject and fails to understand the freedom of consciousness, fails to articulate how we think and how we do anything: the realities of thought and action. Science cannot see the data before it, such data fail to fit its metaphysical model. **Axiom 12** then **Autonomism is an empirical truth we verify in our being conscious**.

So to the emptiness of consciousness, again from **IC&VM** Block 2 Cell 6:- ***This brings the argument to the point of its unitary treatment of autonomous, intelligent life. The philosophical pursuit of what is the self misses the self-evident, intuited truth of our situation. The philosophical enquiry requires some content for the self, empirically some bundle of experience, metaphysically some necessary something we know not what presupposing divine attributes beyond reason -souls-. Instead what we have undeniably from introspection is a will o' the wisp emptiness, something that watches our watching, always there, our self-consciousness. It is spatially and temporally determinable and spatially distinguishable from all other things. But it is not identifiable as a physical thing; very roughly it is like the way light is a property of spontaneously combusting methane -the will o' the wisp-. It enables us to stand back from what we are. It undermines that we are what we are. It does not require a history. All memory may be lost but it persists. If memory is lost, it is the same as it was previously, but it does not know this, unless inferentially constructed. Because its truth is its emptiness it can have no identity beyond the particular, 'spontaneously combusting', physical thing to which it belongs. It can have no after-life, cannot reincarnate. It is us but not us. It is why we are free. Some of us have an incoherent fear that after death one might begin again in another body but without any memory from any past life, without any means of knowing this had happened to one, and the unfairness of this seems that one might have to indefinitely suffer what life is, over and over again, but the emptiness of self-consciousness shows how incoherent this fear is. Self-consciousness is the property of particular bodies. Each time it has a unique numerical identity. Each time it is the same thing but a different thing. Self-consciousness is unique to the individual but it is nothing, just the awareness of*** *awareness, but it makes all the difference. You are your phenomenal self, but you can never be at one with this, i.e. the awareness of this. This self-reflexivity is not though some separate faculty, like some genie released from the lamp, instead it is part and parcel of our kind of experience, it is always at one with the subject of our bidding and not its object. These then are some of the remarks needed to understand what it is that self-consciousness adds to consciousness. We might not have it, we might just be conscious - some of us might be so constituted: the arbitrary variations of living things-. In a unitary treatment of autonomous life the aspect of emptiness has to be set alongside that which is just given, and so determined. There is no denial in any of this of a material, causal order. Clearly most of us just find ourselves unable to jump two metres high, and similarly, and more unified, more integrated into consciousness, we find ourselves, e.g. unable to remember the name of Socrates' lover, or calculate in our heads an odd six figure number divided by an even three figure number. The unity of ourselves is ourselves looking on at ourselves. The given and the unmade.*

What is being said is that there can be a loss of everything that a particular consciousness is conscious of, although it must be that at such a loss some given residues which give a certain complexion to any particular consciousness will remain, but that that particular consciousness persists and will begin some reconstruction. And this raises questions about the beginnings of life and its consciousness and the development of this consciousness, but what might be called genetic epistemology is beyond the competence of this investigation. Instead, what has to be said, is that there must be a bare identity to any particular consciousness, certainly something replete in the midst of life and replete despite the contingency of content. In this regard empty consciousness is not just a registering, it is thought and feeling, a contentless straining towards the world. It is our vitalism I tried to bring to life in a poem, a poem entitled **Touch.**

TOUCH

Phrase used was *turning against flesh*.
Perhaps meaning then, losing touch?
Intimacy starts life touching,
Cut to the quick, quivering fresh.
Slipping away we strain to clutch,
New we are sucking and clutching.

Flesh is sense of touch transitive
And this is flesh being alive,
The root of what is consciousness.
To cloak flesh is palliative,
Shrouded, nakedness can't survive;
Being touched dies in holiness.

There is not awareness and touch,
Touch is physical awareness.
Paradise is not physical
Though pilgrims bear pain of the crutch,
While Christ's body spreads tenderness.
Clapping life is not spiritual!

Seeking pressure of calf and thigh,
Contiguous, sweating pleasures
Raging through a lusting body,
Sound in the ear, sight in the eye,
Distant, not intimate measures
Of all life pounding heartily.

Harvey's capering bloody point
Shrank away from an unkind probe,
Reich's bions are erogenous.

**Moments to live moments appoint
Vitalism's truth to disrobe
Dead mechanism's causal rust.**

**Only in a living body
Does dreaming make a kind of sense,
But sleep in Imitation Games
Is no more than a parody,
Background in computer science
Never is sweet, nocturnal flames.**

**It may be smartly argued that
Events still flow through processors
When all the outputs hibernate,
But a dream is as if it's at
And how can our simulators
Fool themselves that they masturbate?**

**Living machines would have to be
Alive with full tactile feeling,
And deceived in dreaming feeling.
Flesh, grounds of our reality.
Warm duvet as we leave sleeping
Melts, dissolves dreamt believing.**

**Compare background and foreground tasks
And inputs, outputs and display
To simulate the world of touch,
Mechanical hands grope to clasp
Gather data bytes to convey
So like, yet do not lust as such.**

Pinocchio's **strings like piped programmes
Raise him up to resemble life
An image that's identical,
But to be real he needs some drams,
Fairy spirit, a kiss mid-life,
Wood transfused to flesh. Miracle!**

EXPOSITION continued

And this emptiness of consciousness entails non-representational thought. We have thought without language, we think without words or representations, this the reality of thought and consciousness, and something theory has lost touch with, something obvious we do not know about ourselves. Much more is to be made of this later. For now it can be remarked that structuralism et al contains this loss of touch. Of course there is empirical and cultural data about language. So it might seem we can undermine the authenticity of thought. It might seem all our thinking conducted through representation simply utilises historically and culturally pre-formed building blocks so that, in a way, we repeat ad nauseam a predetermined narrative. The idea that we are trapped in a logic and ideological conventions and sound bytes, which leave thought entirely determined by systems (many of which are accidental or fortuitously, for those they benefit, unintentional) which control our economic lives, misconstrues thinking-consciousness, because thinking is independent of those representational or linguistic systems, whose intentional function is the communication of thought (roughly and not the other way round, i.e. not more complete than thought itself). To **think** otherwise (self-refuting in itself) is the **structuralist fallacy**. By default structuralism describes computers (existing) and in intent reifies living thought, and of course is open to the contradiction of a transcendentalism which its premises deny. Consciousness necessarily is revolutionary. **Axiom 13 An empty conscious-**

ness is a thinking consciousness.

So to move to a summation of the bits already discussed.

Axiom 1 A theory of consciousness has to allow that consciousness could be created artificially.

Axiom 2 Consciousness is to be treated as a self evident, spontaneously emergent, physical property of some physical particulars.

Axiom 3 Consciousness holds within it a living privacy.

Axiom 4 Mind is a superstructure generated by and resting on a base of self-consciousness.

Axiom 5 Consciousness is a physical property of neurobiology but neurobiological facts are not a translation of conscious life.

Axiom 6 A self-conscious consciousness continually confronts the choice of divulging or withholding.

Axiom 7 Consciousness is broad, evolving, stretching and always fully engaged in a struggle with reification, where, in the first place, it finds itself.

Axiom 8 Second lives and madness are the unforeseeable freedom of self-consciousness.

Axiom 9 Free conscious beings can choose otherwise than what they have chosen.

Axiom 10 The making of pain in a machine requires a multi-layered, transparent, self-reflexive and transitive depth of genuine awareness.

Axiom 11 Consciousness is an accidental fact of physical existence.

Axiom 12 Autonomism is an empirical truth we verify in our being conscious.

Axiom 13 An empty consciousness is a thinking consciousness.

DIRECT PERCEPTION

Addressing perception is the standard starting point for much theoretical grappling with mind and consciousness. It is where, very recently, Honderich starts his **Actual Consciousness**. A long time ago it is where **Kant** starts the **Critique of Pure Reason**. Partly it is a piece of empiricist dogma; the mind being a tabula rasa, without knowledge, until experience -perception- begins to write on it or fill it. But even for those who wish to begin their certainties with apriori first principles this only takes place after an examination of what are called illusions of perception -deceit of the senses-.

It is though a good place to start, not for these historical reasons, but because it is where early errors begin and grow. The problem is grasping perception as it is rather than trying to explain it away as something which it is not. This is not to concede that perception has its own isolated, temporal moments in consciousness. Whatever is to be said about perception must integrate it into the totality of consciousness at all times. There is never **just** perception.

In **IC&VM** Block 2 Cell 6 I mount a sustained argument against the account of perception given in **Francis Crick**'s book **The Astonishing Hypothesis**. The astonishing hypothesis is that '***you are nothing but a pack of neurons***' and the account of perception in Crick's words is that '***what one sees is a symbolic interpretation of the world***', which '***seems so like* the real thing**' but '***in fact we have no direct knowledge of objects in the world ....Seeing involves***

***active processes in your brain that lead to an explicit, multi-level, symbolic interpretation of the visual scene*'!**

IC&VM sets against these generalities (absurd) a different generality. '***The eye is not just an instrument, a means, it is alive, full of consciousness, and is coming into the world. It greets, it looks at, and in turn is greeted and looked at. With our eyes we make contact and encounter the world of others. The eye acts on the world, it comes towards one, it comes into oneself, it is not just the passive recipient of light as is the camera. The eye forbids, coaxes, twinkles, gleams, seethes, melts, shares: communicates. These are all things of experience, the experience of consciousness, something intelligible in itself, and so for millennia, unproblematic. It is philosophical scepticism, the unrecognised parasite in the body of scientific observation, which drags science into a cul-de-sac from which it risks, paradoxically, its own sanity, because, in its own terms, science cannot be sure that the cul-de-sac is not solipsistic.***'

These generalities are a move towards being in the right place to focus on the plausibility of direct perception. Crick blurs this plausibility basically on two grounds. One, arguments from illusion, the other the story of the physical mechanics of perception. Both of these are dealt with extensively in **IC&VM**, and will not be repeated here.

Sensing is always fully loaded with all we are. It is never just light waves being mechanically reformulated in a camera obscura. As it was for Samuel Palmer it is never just the moon, the elm and its shadow cast upon a wall, it is always as it was for Palmer, replete with being, with feeling, desire, thought and all of these always interlinked, integrated, not things to be isolated the one from the other, but always suffused with each other. Real experience. And it is always reaching out, making contact. This is how we talk and think about sensing. We make contact with the world both near and far, and sometimes the nearness is what is intimately ourselves. We are alive and in the midst of life, pulsating in our contact with the world. We are not outside our world visited by phantasmagoria about which we weave some hypothesis concerning external causation which we will never be in a position to verify. This point of view is the bloodless position of an objective science that supposes for a moment that it has, despite its premises, the omniscience of a deus ex machina. What, because of intellectual transcendentalism, it fails to make contact with is the empirical certainties of active, living beings in their actual life. So we do sense (e.g. see, feel, hear) all the things there are, where they are and when they are, and these are not two areas remote from each other but instead there is constant interaction between the two, things of the world and active living beings. And, who would deny this apart from a few scientists trying to reduce the self-evidently intelligible to a narrowly conceived physical substratum, who tell how our world is entirely different from our lived experience and how we sense this.

This is not to say the questions raised by science about sense should not be faced, and in large measure this I have done already in **IC&VM**. The main attempt here, however, concerns directing attention to the plausibility of our normal discourse and belief. A matter of reminding of how we absorb unguarded experience and how impossible it is to live, moment to moment, with Crick-like reinterpretation of our instantaneous involvement in the medium of our existence, whatever our intellectual convictions.

There are questions about what we take to be real and the length of time it takes for light waves to travel, or the specifics of optical structure determining the colours we see, or the delusions of sense that cannot possibly match the reality of cause. But then any intellectual reflex that hastily, on the basis of these questions, removes us from our intimate, conjugal media fails to imagine other narratives.

How could seeing be other than time-lapsed. It is dependent on an unceasing tsunami of light, the medium in which we discover the objects of perception. There has to be a spatial gap between ourselves as perceivers and the things we see. A coherent concept of direct seeing cannot require the obliteration of any such gap. What we see is mostly almost instantaneous with how what we see is, but because our world is a physical world what we see is always in a sense what was, on a continuum from almost terrestrially instantaneous to cosmically lapsed (remembering it is a long-standing law of physics, although itself challenged from time to time, that nothing travels faster than the speed of light). Is this a problem? Even when time-lapse is of greater magnitude we make allowances so as not to be misled. But let us try affirming what we live by, namely, what is seen is seen directly, so not just something in our heads and posing difficulties about the difference between what might be there and some mental construct. For the most part what drives such scientifically created paradox is really a version of the **Zeno** paradoxes, ably discredited a long time ago. Of course theoretically there is the logical possibility of a gap but this does not prevent us crossing the gap and so not standing in need of those '*mind the gap*' notifications. If we want to be pedantic the seagull we see continuously, through micro-seconds, hovering above the verdant valley as we look out from our own personal eerie was where we see it some micro-second before seen, **but that is not to say that we do not see it out there above the valley in the microsecond it was there,** and so on continuously. This is seeing. The gap does not mean we can't see it. Clearly we do and how could anyone live continuously without accepting the immediacy of this. **To demand that for direct sensing to be possible there must be no gap is to posit a world without physicality and so deny the premises of science: so a contradiction.** And even on the alternative, 'scientific' accounts there will be gaps even when all the processes have been gathered inside the head, so ipso facto tempting the same epistemological confusions. There will be micro-seconds between an image event on the retina and the formation of the supposed symbolic interpretation of the cognitive moment. If we are to heed the gap all we are left with are spontaneously-arising, deserted propositions and a cyber solipsism. This is a negation of science. And, in ranting mood, how can science push us to symbolic interpretation anyway, as though seeing is the cognition of a verbal proposition, when in fact seeing is an immersion in the visual world, the discovery of objects through the light they emit, no words/symbols need/have to be involved. We continuously locate, confront, discover the physical world through direct visual contact. And the thought of an epistemological gap because of spatial gaps is a red-herring. In sensing we know the world through touch. Deprived of vision we can directly experience the world through touch where there is no gap apart from a contrived episodic reading of what is continuous, neural process, and objecting on these grounds is demanding logically impossible criteria for **directness**. I see the hillside and feel the ground on which I stand, all these words meaning what they say.

Colour perception varies from species to species and between members of species, dependent on the nature of perceptors or so science tells us, but this does not make colour something mental, i.e. the traditional, philosophical mistake of secondary qualities. Instead the world we sense has many appearances. This is its reality. **If** I see the adjacent hillside as green in colour and another sees it red, these are appearances the hillside exhibits or could. The hillside could be both red and green, and which of these appearances one sees would depend upon our sensory sensitivities. At the same time this is not to confound distinctions between sensing and hallucinating. Hallucinations are possible too, drug induced or when in a fever, or taken over by a psychotic episode, the sense that we see something when we don't, rather like dreaming. Equally the possibility of hallucination does not preclude the possibility of a hillside which has both the appearance of being red and being green, and, accordingly, can be directly seen as both a red and green hillside, which the hillside would be if genuinely seen as being both by differently constituted sensory sensitivities. I was brought up in a philosophical school which held that it was a necessary truth that an object could not be both red and green all over. Such a proposition was held up as a paradigm of propositions called *analytic*: somehow necessarily true because of the meaning of their terms. The present reflections deny strict analyticity to such propositions. Analyticity has to be modified to accommodate what might seem to be contradictory appearances. A red and green hillside would have to be accepted and treated like the duck/rabbit figure. Although if your sensory sensitivity is to see a green hillside then obviously for you the hillside cannot be seen as red and green (all over): a logical truth founded on a contingent fact, like a bachelor being an unmarried man being founded on a fact of language.

Illusions, however, present a subtly different case. With the duck/rabbit there is an x (say marks on paper, chalk on a board etc.) which has both the appearance of a duck and the appearance of a rabbit. This case is not the same as the colour of the hillside. X is not a rabbit nor a duck, instead we say it looks like a duck or it looks like a rabbit, but it does look like both, although it cannot be seen as both simultaneously. The posited hillside on the other hand is red. It is also green. It may be other colours as well. At the same time the greenness of the hill is one of the hills appearances. The hillside is green. It is not that it looks like it is green but is not green. Illusions on the other hand have to be treated differently. Looking at the duck/rabbit we know we are looking neither at a duck nor a rabbit. Illusions on the other hand delude. However, rather than say that what we see, in such a case, is some *symbolic interpretation of the visual scene*, the more direct way of dealing with the case is to accept what straightforwardly we think, namely, that what we see directly is a delusive appearance. There is no reason to rule out the possibility of reality containing appearances which are delusive. The fact that we may be deluded should not automatically make us think that we never see anything directly. What we see, sometimes directly is an appearance which is delusive. Thus, for example, with what is called the *spreading-effect* the actual shade, tone, colour of a particular visually isolated pigment cannot be seen as such in juxtaposition with other specific pigments but is transformed by the juxtaposition. The reality presents a delusive appearance and this is what we see, although we have means whereby we rid ourselves of the delusion and see through the illusion. This is the case where what we see looks like, say, a lighter shade of blue than is actually there, but not so as to allow us to say that it is not that shade, as we will say of the duck/rabbit that 'it is not a rabbit'. This is to say we can see it is not a rabbit but cannot see the blue is a darker shade than it looks. The look that it has deludes and this is what we see, the delusive appearance. The red and the green putative hillside is not delusive. Whatever our sensory sensitivity we cannot see the darker shade of blue when the spreading-effect operates. None of these cases, however, should drive us away from what we **know** to be the case, namely that all of these things we see and see directly'. We do commune with the reality in which we are situated.

It is absolutely essential that we do not get dissuaded from normal discourse and persuasions. This is to say it is essential we trust our **empirically** founded commonplaces. The intellectual challenge to this, sort of, begins with Platonism. It is the construction of an elite who attempt to impose ignorance on (and this may seem inflated but is in fact commonplace) a threatening, lumpen class so as to better secure its own power. Ordinary understanding can no longer be trusted. There may be no objective order independent of mind, and all that mind can be aware of are shadows and theories about how mind creates what we are deluded into supposing is reality, but which is only shadow and as such requires an elite of shadow-readers. Immediately labour (the producer of use-value) is robbed of its known world (the way of imposing control). This has to be redressed. However, despite the various manifestations of intellectual conceit, normal discourse and persuasions do remain intact, and remain in force, strangely, for everyone (regardless of class) in their ordinary dealings with the world.

So what are the implications of what is being called normal discourse? What is it we need to be reminding ourselves about not to be diverted from self-evident but non-critical knowledge? And how is normal discourse about direct sense to be saved from mocking accusation. A machine, lacking consciousness, receiving images from a world outside itself will apply programmes to those images and so generate propositions about them and by implication about the world. However not being conscious it will not **think** it **sees** what its images are images of. This will be a dimension absent from its activity. Conscious beings have no hesitation about this. **Seeing** x is contacting x directly. X is actually there and seeing it is discovering this actuality directly. There is no gap, to be bridged by representations and interpreted symbolism. The atrocity, the erotic revelation, the jug of water, the sea, the laptop screen etc., are located where they are, by seeing, by sensing. This is the reality of seeing, it is more than mechanical causation, it is living sense. The reality of seeing is a property of some mechanical causation, but is something over and above this causation: it is the spark of life, it is conscious seeing, transforming the causation.

In the cabin where I think and write there is a reproduction of the Gioconda pinned to the inside of the door. Often I turn to it and look. The reality is a two-dimensional surface looking like a three-dimensional world, a figure set against a landscape. This reality I see. I look through the space in front of it until my gaze comes to rest, perhaps on the eyes, or the hands or the enigmatic smile or the balance of the whole (all of these part of the representational mode). It is in front of me and there is no difficulty in making direct contact with it. There is nothing standing in the way. My gaze falls on the object. I see it. And I see it where it is, the reproduction pinned to the door, a pinning I affected. My world is three-dimensional and my eyes allow me to enter this world and locate its objects and beings, and if I am blind I cannot do this in the same way. If I am blind I do not see the things of the world. I accept that seeing is possible because light waves reach my eyes, but I do not see light waves or retinal images caused by light waves, instead by means of light I directly see the object from which its light comes. **And we all behave as if this is true, for no other reason than that it is true.** At some point one of the scientific stories invokes the concept of seeing with all the force I am attaching to it but makes this seeing take place inside the head, but if you can see e.g. the retinal image, why can you not see a three dimensional world. In both cases seeing bridges a gap. This is what seeing is. Of course another scientific story gets rid of seeing altogether whereby, *probably,* an external cause triggers a sequence of computational relations, neural, symbolic, semantic, issuing the behaviour, the speech act, 'I see the Giaconda pinned to the cabin door', and so what vulgar belief holds to be the experience of directly seeing the reproduction does not take place at all according to this reductive interpretation. But what is the compulsion of this assertion? It is that science cannot reduce direct seeing to its physical terms and so has to deny its existence, whereas the plain truth is that direct seeing is itself an irreducible, physical process. Ordinary language, evolutionarily selected, is already fundamental, and what it refers to runs alongside those other physical processes which science is competent to describe and explain. Metaphysics, typically, is an error in moving from the logical to the ontological, and metaphysically engendered ontological claims not only conjure rabbits from hats but also put ladies into boxes and then make them disappear. None of this is objectionable if confined to **as ifness** (compare **Beyond Art** and **Invisible Cells & Vanishing Masses**) but trouble begins when claiming ontological truth. The strange thing is that in claiming that no one can have direct knowledge of the Gioconda on the door but only have *a multi-level, symbolic interpretation of the visual scene* it is not denied that we all have the conviction that we see the Gioconda directly, even those of us who claim to know that we don't. You would think this might give one pause to re-examine the act of seeing/looking and so accord it a different status. And anyhow what on earth is the concept of '*the visual scene*' doing in this exposition of so-called objective process. There is a terrible sloppiness to this putatively robust, scientific rigour.

Think of what it is like to be in the world doing something: any doing is always multi-faceted. The idea of '*multi-level symbolic interpretation*' is a general assertion, not just about a singular, perceptual moment, but about our whole sensory being, in which, it is alleged, we never have direct knowledge of any of the things we wrongly suppose we interact with. So think of a bowler, bowling a ball, in cricket. The bowler stands at his mark, knowing this, and surveys the scene, looking around. The ground surrounds the bowler, spreads out from him, neat and trim, an intense, smooth green. The bowler occupies and absorbs his location, and directly in front there is the rolled, flat strip of the wicket, a somewhat parched ochre contrasting with the green outfield; it reveals a few cracks from wear and tear and targeted vision. The bowler checks his fielders are correctly placed, gestures to the right for the mid-on to assume a slightly more forward position, and then, grasping the shiny, red ball, the touch confirming the forefinger has good contact with the stitched seam, begins to trundle in through the breeze and warm sunshine, towards the batsman guarding the wicket at the other end, with the wicket-keeper behind seen to be crouched and ready. The speed of the bowler's run increases and this is sensed both in muscle and bone and visually as the distance between the start of the run and the release point is seen to rapidly decrease. The ball leaves the hand and the bowler watches its shifting trajectory bending through the air towards the batsman and the wicket defended, until it pitches with a thud, arcs to the left, eluding the batsman's forward defensive and clipping the off-stump, leaving it askew and the bails flying through the air. All of this, what else is there to say, is experienced directly by the bowler in bowling, in being on the cricket pitch, alive and interacting with the totality of the village-green. Ask the bowler after he has taken the wicket whether all of this is just something symbolic in the mind and that none of it amounts to having direct intercourse with objects and persons in the world. To think this is nonsense. To take issue in this way is not some reversion to folk-philosophy, rather it is an accusation that science fails to recognise the ever-present dangers of Ptolemaic certainty. Seeing is contacting something external, a way of obtaining it. Perception is obtaining something. It is not 'reflected' (?) data, in a box, translated into symbols. The latter is what a computer (existing) can do, and the information content may not differ radically from what something that **sees** reports, but these are different means to an end, although seeing is a more diffuse process and not singularly driven to such ends. The computational system does not **see** unless it is **conscious**. And it is not conscious just because it simulates the outputs of conscious beings. And it is not being denied that computational systems simulate. They are Imitation Game machines.

Physics claims the high ground in explaining perception, presenting itself as precise, careful, meticulous. It can measure frequencies of the electromagnetic spectrum and what it calls visible light consists of wavelengths ranging from about 780 nanometer down to 390 nanometer (applying a precise system of measurement). This is physics. This is its own field. It then tells a story about light waves **striking** the eye, which quickly becomes a biological story, so borrowing from a related science. Light enters the eye through the pupil, striking the retina, which is lined with rods (intensity sensors) and cones (wavelength sensors), causing a chemical (another science is invoked) reaction that in turn causes electrical impulses being sent along nerves to the brain. This the story of perception according to physics. It does not require denial. Clearly a physical process takes place and is describable, and this may be the description of it. Physics textbooks are assertively confident in telling how the world is, but on examination their **whole** story reveals glaring imprecision in thought and language, and, if this is so, we should be hesitant accepting that due care has gone into its parts. To highlight some imprecision. A typical physics textbook takes the story of perception forward in supposing that white light '*is incident upon the retina*'. '*Upon striking the retina, the physiological occurs: photochemical reactions occur within the cones to produce electrical impulses that are sent along nerves to the brain. The cones respond to the incident light by sending a message forward to the brain, saying "Light is hitting me". Upon reaching the brain, the psychological occurs: the brain detects the electrical message being sent by the cones and interprets the messages. The brain responds by saying "it is white".*' Wow! Are we not supposed to notice a traversing of modes, from hard science to Disney? Is this a poorly executed conjuring trick performed in such a way as to try to make us think that it is easy to move from a measureable, physical narrative to a conscious, cognitive narrative of what we call **seeing**? Once we have electrical impulses **saying** '*Light is hitting me*' and brains **saying** '*it is white*' fantasy has already bridged the gap to human perception. The question of how we move from electrical impulse to the linguistic proposition '*Light is hitting me*', is an enormous question. Not something that can be bypassed by any easy assumption that nothing strange is going on. The question, in itself, should be the attempted beginnings of a science, not a sleight of hand to avoid us thinking. Starkly the point is that electrical impulses do not **say** '*Light is hitting me*'. Brains do not **say** '*it is white*'. They don't **say**. Conscious beings say, because they are conscious, saying is more than the occurrence of a sequence of symbols, not that brains and electrical impulses manage even that. The difficulty here is that of being able to distinguish between physical process being all that conscious thought is and physical process becoming conscious thought. These are distinct problems. One for a reductive science and the other for my undertaking. Science will have, or is determined to get, all the quantified, measured and so physical data of perception and this equated with what seeing is. The problem is how to move plausibly from the data to seeing and saying, and to claim neither is diminished by the assertion of equivalence. And the problem is similar from the other end. Thus how is seeing directly a physical property of a quantifiable physical process but not itself reducible to the physical data? How can it be physical but transcend current physics? There are two languages here and the relations between them unclear. But this we have to concentrate on in order to tell a better story than the ones available to us.

Physics tells a loose, imprecise story but masquerading as tight and precise. It says to be technically appropriate that colour is just a physiological and psychological response to the wavelengths of light entering the eye. But this conclusion rests on what physics confesses it does not understand, or is willing to fudge over. So physics says when the brain recognizes that messages are being sent by all three cones it '***somehow interprets*** *this to mean that white light has entered the eye*', and that when the brain recognizes that the light has activated both the red and green cones it '***somehow interprets*** *this to mean that the object is yellow*'. These are typical formulations of a position within physics textbooks. At no point is the conclusion baulked because it rests on 'somehow interprets', even though this is left a mystery at the heart of physics, but on which firm conclusions are reached about what perception is. Nor is the slippage from the brain interpreting '*white light has entered the eye*' to '*the object is yellow*' apparently any cause for concern. This is where the difference between standards of argument in physics and philosophy reflect on the pursuit of truth. '*White light has entered the eye*' and '*the object is yellow*' are propositions of a different order, purported facts about different objects. Arbitrarily the brain is supposed to throw out this difference without any kind of spasm, although if the brain was any kind of computer it would experience Boolean hesitations. And anyhow brains do no such things whereas perceivers do. Physics knows that what it proposes, what its methodology remorselessly has led it to, generally would be thought of as preposterous. So we find '*..to be technically appropriate, a person would refer to* ***yellow light*** *as* ***light that creates a yellow appearance****. Yet, to maintain a larger collection of friendships, a person would refer to* ***yellow light*** *as* ***yellow light****.*'!! Of course what '*light that creates a yellow appearance*' means in this context is light that creates the appearance of an object that is yellow. What it might be for an object to be yellow is not contemplated, in fact what physics is claiming is that objects are without colour. Apparently the existence of yellow is just something the brain **says**. But on this analysis there are further questions. What is yellow? Why is there yellow? What is this brain up to, inventing experiences of what does not exist? And if yellow is a content the brain finds, something distinct, **yellowness**, why could this (apriori) not be a property of things? Is it theoretical nonsense to imagine a world containing things that are yellow? Physics attempts to cast doubt on the objectivity of yellow things by pointing out that we cannot distinguish between a surface illuminated by a yellow (whatever this is) spotlight and one illuminated by overlapping red and green (again whatever they are) spotlights. Despite the causal difference we will perceive both these surfaces as having a yellow appearance. The conclusion is that yellow must therefore be just a peculiarity of subjectivities. So, '*In a technical sense, it is really not appropriate to refer to light as being colored. Light is simply a wave with a specific wavelength or a mixture of wavelengths; it has no color in and of itself. An object that is emitting or reflecting light to our eye appears to have a specific color as the result of the eye-brain response to wavelength.*'

It is necessary to reflect on how preposterous this narrative is. The world is without colour. But how the world actually is we cannot perceive because of subjective contingencies, although science enables us to accurately describe this reality through nanometer readings, but we have no unmediated direct experience of it (Platonism?). Light of different frequencies strike the eye and then the eye and brain talk to each other (well send messages at least) to construct a set of parallel, invented distinctions of colour (which sort of appear from nowhere) corresponding to the wavelength differentiations, and this physical process, containing on some **unexplained** basis cognitive elements, then, again **unexplained**, becomes conscious perception (the experience, for example, of yellow things). We are locked in an inner world of colour which cannot become an objective order independent of consciousness. Despite this supposed divide between the objective and subjective orders science spans the divide to explain both!

But what can be set against this preposterous, quasi- solipsistic story, i.e. reality according to physics. Well a world of coloured things **actually** seen. A reality we all accept in routine behaviour, science-convinced or otherwise. The physical story of lightwaves, retinas, cones and brains stays as it is. It does not need disputing. But this physical process is the process of seeing. We do not somehow see things which are in the mind. We see things whose reality is independent of mind. But if it can be allowed we see things in the mind, then what is allowed is a bridging of a gap, and if this is theoretically possible then in looking at mind-independent things why should we not bridge the gap to those things looked at. Again, this is what seeing is, the reality of seeing; a concept standard physics will not grasp (perhaps because it is a slave to a reality of cause and contiguity). Seeing is raw, irreducible, physical process, something that happens, happens in the physical world, as basic as the frequencies of the electromagnetic spectrum. But can this be so when yellow spotlights cause the same yellow appearance as red and green overlapping spots. Doesn't the explanation of this, point to some limitation or deficiency in our subjectivity which prevents us from distinguishing this objective difference in reality, confining us in delusion? But the reality according to physics is not yellow spots and overlapping red and green spots but three specific nanometer frequencies, and why shouldn't overlapping frequencies produce the same effect as a singular frequency, just different ways in which an illumined surface is yellow in colour and seen as such. And to face the alleged contradiction head on why shouldn't projected red and green spots produce the same coloured illumination as a projected yellow spot. We have only to use our eyes to see that they do. This is how the world is. In fact the world is more complex than this. Objects have different appearances and our tuning enables us to see some of these and not others. It is not that our tuning manufactures sense data that we then, precariously, ascribe to the world, rather it is that the world makes its various appearances available to our tunings. There is nothing preposterous about this, it is just that what we are able to see depends upon our vantage point. How difficult is that to grasp? If too far away you can't see what is written, but closer you can, this doesn't mean nothing is written.

All the points made here can be generalised. They are points about the **senses**. Seeing, hearing, smelling, tasting, feeling. The physical has various properties, some of which are visual, others auditory, olfactory, gustatory, tactile, sensory. Consciousness is the means whereby these properties are located and known. This is physical communing. All that is physical is constantly interacting and consciousness is part of this interaction. Common-sense would say of 'I see the wall' that the wall is there and that I see it as it is and where it is: it is there and I am here and here I see it there. The only departure from common-sense in my account (not really a departure but more a filling in of what is not considered) is that **seeing** is part of the physical order, and more particularly a physical part of conscious, physical things. Science, on the other hand, wishes to ridicule common-sense consigning it to an understanding from the dark-ages, despite the scientific cognoscenti behaving conventionally in accordance with common-sense in everything (e.g. masturbating), apart that is when in professional mode and donning sceptical masks.

But to consider the issues more deeply, or, at least, more philosophically (often a red-rag in itself to the scientific community). Is perception as construed by, what is being called, common-sense, possible, we might say logically possible? Is the concept logically incoherent? Is it a necessary truth that any object emitting light to what we may call a percipient can only be known as a representation within the percipient i.e. cannot be known directly? But if the answer to this is 'yes', then it is a reductio absurdum. The pertinent question is how do we know the representation? Is there a representation within us that we apprehend and interpret? But how would we do that? The critical distinction has not changed, there is a percipient and something to perceive, and science cannot grasp how this is possible, but in the case of the representation it can, yet how is this different? Or is it the case (and I would not want to fathom the intricacies and depths that might be needed to make this transparent) that the representation is itself, knowing the object, but if that is so, why is this not seeing the object directly, i.e. what happens in the individual when an object is seen. Nor is it any use to argue that the mind contains rules and conventions for interpreting visual, sense-data as linguistic propositions, as is sort of mirrored in the workings of computers. The computer's camera records the wall and the linguistic programme issues a linguistic proposition on screen, or anywhere you like it, which is, 'there is a wall'. Well yes, that can happen or rather does, but where is consciousness? 'There is a wall' i.e. text on a screen is not seeing the wall, is not in itself conscious; something very large is missing. There are those though who will argue that there is no more to perceptual consciousness than a visual representation decoded into linguistic propositions which can cover all the relevant questions about what the visual representation is a representation of, cognitively indistinguishable from the answers that any alleged percipient would provide. The strong claim following from this is that computers with this facility are perceptually conscious, and the stronger claim being that this is all there is to the question of consciousness. Honderich in his ***Actual Consciousness*** argues that it is indubitable that human perceivers are conscious but that it is much less certain the same can be said of computers (those having this imitative performance), and so in the case of computers (as they are) something clearly is missing. But this is a weak claim. It is indubitable that the specified computers are not conscious. Convincingly imitating the behaviour of x is not being x. None of us treat computers as conscious things. We do treat them as things which can do some of what conscious things can do, and where they can they can even do whatever it is better (quicker, more accurately). But pursuing this argument is difficult. It is tempting to keep pointing to responses that conscious things can and do make which given computers do not make but programming additions can always counter this. I can ask a percipient if looking at the white wall is boring and the percipient will provide a relevant answer, whereas asking my computer the same questions with its camera picking up the white wall just leads to a malfunction of, or some mis-classifications in its interrogative programmes. But a programmer can enable the computer to transcend this, so that the next time the question is asked, a relevant answer is provided (*'I do not find it boring Rog, instead I list the similarities between this and a Malevic'*), well, anyway, this line of text appears on the screen. It is not at the imitative level we discover what is missing. What is missing is life and its recognition is basic. The computer is not alive and we all know this, and so as consequences it neither sees nor thinks nor is conscious, although to reaffirm the point made many times before, this is not to say a machine could not come to life. This is the kind of materialism framing this discussion. **Life** is not though the end of the matter. It is itself complex. A growing blade of grass has life, is alive. This is what we say but we would be hard-pressed to allow it more affinity with a conscious thing than a computer. But life is a much stronger candidate for what's missing than say feeling. In his **Guardian** review of **Harari's** ***Homo Deus***, **David Runciman** makes feeling the defining attribute of consciousness but this is much too simplistic even though affective consciousness has to be integrated into consciousness in general. There must be passages of consciousness where in the normal understanding of feeling, feeling is absent. Not so though with life, or at least life and human consciousness, and, of course the evidence for life with humans often is the presence of feelings. It is the variety of the living response to viewing a white wall that highlights what's missing, and any programmer attempting to overcome this difference (attempting to imitate or make consciousness) would have to stretch and strain to approximate. A human viewer of the wall is open to an infinite range of subjective responses to the viewing over and above the straightforward cognitive report of what is confronted, and, very importantly, all of them genuine. So if the observer is desperate to leave viewing the wall because he or she really wants to look out the window but can't because of having to follow orders, and feels frustrated as a consequence, how can any existing computer genuinely be in this position, ie. desperate, wanting, frustrated. This is life and this is what's missing.

Consciousness is life in physical contact with the world and so not a camera obscura or a Platonic cave, not something one removed from reality, not a virtuality, although it entertains the virtual or is its source. Materialism is not metaphysics. Science dismisses philosophy yet is not aware that covert, philosophical premises contaminate its self-congratulatory empiricism and lead it to being lost in a latent metaphysics.

But does a concept of direct seeing fail to acknowledge the subjectivities of perception and is it not the achievement of the science of perception to draw attention to them? There has been some discussion of colour perception and some argument to set aside the problems in favour of a world exhibiting a multiplicity of appearances and so avoiding the difficulties of contradiction. But is this strong enough? Consider a different but analogous case. Namely the problem of myopia. From the point of view of direct seeing we will say that a person with normal vision looks at the world and sees it directly as it is, and unmediated. But then what does a short-sighted person see. For this percipient the world is a blur, nothing sharply defined, much of the detail missing. Is this perceptual experience not a product of subjective circumstance and so, if this is so, can we not attribute the nature of normal perception to a difference in subjective constitution? And can we not add to these difficulties by raising the case of looking at the world through a coloured lens or glass or revert to a more head-on consideration of colour blindness? Do these cases not give rise to difficulties? To what extent can we say that the world is as it is experienced by a short-sighted person, and of course there are many degrees of short-sightedness and so, we might say many different worlds to consider? And supposedly the same difficulties beset long-sightedness. Clearly the print in the book is not as it looks to those not long-sighted. What then is seen? So there is a thesis about the world exhibiting a multiplicity of appearances and there must be a thesis about the world being a particular way i.e. not as seen in myopic perception. To say the world **is** as seen by the short-sighted person must be incorrect. But should these cases require a retreat to mediated representations and direct confrontation with our own subjectivities. The choice surely is not between hallucination and something which is the same as an hallucination but externally caused? Compare seeing the Blaskets (islands off S.W. Ireland) on a clear day and through a mist. Things can get in the way of our direct perceptions, some of them subjective, so we have the difference between myopia and looking through coloured glass, or mist or a sandstorm. Here the case is one of not being able to see precisely what is there, but this is not a case of not seeing what is there. 'I saw the Blaskets but not clearly because it was a foggy day.' 'I saw the Blaskets but not clearly because i had lost my glasses.' 'I did not see the Blaskets because although I have normal vision and it was a clear day I can never get beyond my subjective representations of things.' Which is the more absurd? It may be difficult to know what to say about cases like colour-blindness or even double vision but there are, I suspect, many options to consider that retain the directness of perception, before leaping hurriedly to a cul de sac of subjective representation and the solipsism inherent within it, two of which are considered here. Namely, a world of multiple appearances (a duck/rabbit world) and a world in which things get in the way of our directly perceiving as well as we might (drunkenness, twilight, migraine). And these recent instances suggest that there are subjective experiences that are quasi-visual. Optical flashings, 'stars before one's eyes', floaters, but we know these are not things in the world we directly see. We make a distinction. That there maybe some cases where we get confused should not tempt us into the fallacy of a scepticism which argues that because something might be mistaken nothing can ever be known to be correct. This is the well known mistake, although often not heeded, of supposing that the logical possibility of something not being the case is a reason for holding that we can never know that it is the case. This is a case for examining carefully the meanings of 'might'. And there should not be anything surprising or even confusing with the idea that there are external and internal aspects to sensing and that we readily distinguish between them. There is feeling the smoothness of the desk-top with one's hand and there is feeling pins and needles in the hand or an irritation or a tickling sensation. One is a sensation of the world (a direct contact with the world) that our sensing mechanisms make possible, the other a sensation of the sensing mechanisms themselves, and we are usually in no experiential doubt as to which is which. The actual experience of the experiences is fundamentally different. We know that the jagged flashings accompanying migraine headache are not an appearance of a world with which we make direct contact but instead are experiences of the visual sense itself and in no way are evidence for the existence of 'sense-data' or visual representations. The senses of consciousness are alive with their own sensations, the sensations of themselves, and this must be in-keeping with what consciousness is. We see by means of those things that are alive in their own right, like the cat that catches the mouse.

And finally, if the actuality of direct seeing seems hard to grasp perhaps it is because things are one way and not another (although conscious thought is quite the opposite, yet is this way and not another) and the reason for this is that in the end this is what reality is. It just so happens that we can see. The world might have been another way but it isn't. This isn't to deny that there is an evolutionary story to be told about how there is an eye. And if we think about our visual world in its entirety it is all of a piece in being a brute mono-actuality. The medium in which or through which our seeing makes contact with our world is light, and we might say, through the brute fact of light. The mystery of the world being a particular way: light and the consciousness of light. As Einstein admitted, after fifty years of deliberation, he could not say what light quanta were, but they were. Light is composed of photons but photons have no mass, massless particles with energy and momentum. How the physical world is and perception is an integral part of it (how the physical comes to know itself). At some point the world has to be accepted at face value and this has to be the starting point for anything that is science. I said 'finally' but there are so many things going on in this last paragraph that there is an immense subject to be returned to. At the same time something hopelessly muddled has happened if this revisiting suggests constructing some vast, materialist equivalent to **F.H. Bradley**'s **Appearance & Reality**.

NON-REPRESENTATIONAL THOUGHT

From the discussion up to this point emerges the suggestion (really stronger than a suggestion but as there is a long way to go 'suggestion' is more measured) that ordinary discourse about consciousness (founded, obviously, on millennia of lingual practice) is, in itself, at the level of irreducible fundamentals of consciousness. To take an example from the preceding discussion. It suggests we are physical things that see physical things. This is how we talk. Well, to be more precise, rarely do we refer to ourselves as physical things, but we are nonetheless non-controversially physical things, and certainly we speak of ourselves as seeing things (not meaning **things that see** though that too is what we are). Our ordinary discourse assumes there are things to be seen and that we see them directly, that is we see them where and when they are, and there is no suggestion in this that our seeing is mediated by some kind of representation which miraculously we can see when we cannot see things themselves (as they really are). Imposing some disjunction between things to be seen and the seeing of them is the beginnings of a long history of philosophical distortion. A distortion of real and ordinary experience, which should not have infected scientific thought too, but it has, and probably has because science is weak on or dismissive of conceptual analysis.

It is necessary though to ponder the notion of '*irreducible funda-*

mentals of consciousness'. I mean things that are real in their own right and do not require further analysis in order to be understood. Fundamental constituents of our world. At the same level as fundamental, scientific determinations. Alongside scientific determinations. The correction required is both simple and immense, perhaps *sublime* in that older philosophical sense (stretching conceptual frameworks beyond stability). The scope contains at least perception, thought, feeling, desire, intention, action and to reiterate, always fused with each other. What is being said is that matter (a concept still alive in contemporary physics and given that there is anything at all necessarily so, although open to reductions and translations) sees, thinks, feels, desires, intends, acts. Not all of it but certainly (empirically given) some of it. This position when understood refutes any kind of dualism and any concept of disembodied life. This is a materialism and this requires constant repetition. But the implications recast presuppositions about the physical: they enlarge our theoretical grasp of what's physical. Once this is glimpsed there is a possibility of coherence in the idea of releasing some matter from a determinist order. To simplify, when matter thinks, reality is changed, and this is what the universe contains, thinking matter. I am tempted to add clarity here by saying this is dialectical materialism. What the theory describes. For too many there is only the possibility of a reactionary understanding of this. An hegemony is responsible for this, although forcefully countered in my **Invisible Cells & Vanishing Masses**.

Once it is allowed that thought and action are alongside scientifically respected determinants then how can they not be captured by causal chains and so not be part of a determined order? However, if thought and action are physical and are alongside then those other alleged determinants may change, or rather be as they are (i.e. not determinants in the scientifically understood sense), by means of, again to simplify, this time metaphorically, a kind of liberating 'osmosis'. Certainly human experience is varied enough for it not to be easily assimilated into a repeatable order, and it is this that standard science requires if consciousness is to be amenable to standard scientific treatment. Science does not **know** what any of us are going to do. If this is so and if there is an invariable alongside-ness then the certainty of alleged determinants as determinants is much less than secure.

So to make **thought** the subject for a while, i.e. thinking as a physical property. We do not want any false distinctions here. Thought generalises the various aspects of consciousness. **Thinking** is seeing, feeling, desiring, intending, doing and thinking. We might as easily say **being conscious** is seeing, feeling, desiring, intending, doing and thinking. All of this coalesces into the language of consciousness. So many ways to talk about the same thing. A conceptual family of understanding. Seeing the taxi waiting is thinking of it. Feeling anxious about a flight is thinking of what may go wrong. Wanting to chill out is thinking of peaceful times. Intending to visit Ibiza is thinking about the island. Walking across the sands of Casa Blanca is thinking, among other things, of what you are doing. Thinking of Dionysus is thinking. The words need to be nuanced but they are interchangeable. However, standard discussions insist on a tripartite division between, perceptual consciousness, cognitive consciousness and affective consciousness. Starting with perceptual consciousness leads to a concentration on **objects** of consciousness. So, saying we see the moon in a daytime sky sets up the moon, the daytime sky and the one in the other as objects of consciousness. This seems non-controversial. Perhaps this is the **structure** of what it is to be conscious. Those who question direct seeing convert the moon and the sky into representations: these are the objects of consciousness. But whichever way, the view is that consciousness has objects. And so when turning to cognitive consciousness, say, **thinking** of a moon in a daytime sky, it might seem natural that this cognitive consciousness takes as its object some form of representation, of the moon and the sky, perhaps a linguistic representation, say a proposition, or perhaps an image. Similarly with affective consciousness, wanting to see the moon in a daytime sky can be thought of as a longing relationship with a representation of some sort, an image or proposition (it is hard to see how **the longing** itself is a representation unless we concentrate unduly on the word *longing*, but then what would it represent?). Similar requirements are extended to classic non-denotations, the unicorn, the present king of France, etc. But it is precisely fixing on this **structure** that paves the way to being unable to distinguish between consciousness and artificial intelligence.

To try to think differently. When the moon is seen in the daytime sky these are **existing** objects of consciousness. They exist as objects and we are conscious of them. When the moon and the sky are just thought of, say longed for, try to contemplate there being no **existing** objects of consciousness. Try thinking there are no representations, and what are called mental images are nothing at all. Things may *seem like* but that is not being so. That which does not exist does not exist. In no way are there representations as there are paintings in the National Gallery. There are no images as there are projections or digitally activated pictures in your local cinema. None of these things **exist** when we just think of the moon in a daytime sky. But what of lingual representation? Is it not natural to suppose that when we think of something the thinking involves words. Theorists for this position are at least sensitive to lingual thought not taking the form of fully articulated propositions. This much is empirically obvious. It takes time and patience to construct a definitive form of words as in a written proposition for publication. It would be absurd to suggest that this form of words is with one from the very start of whatever thought is involved. Arriving at a definitive proposition requires drafting, modifying, polishing etc. What is in the mind at first surmise cannot be this. But why is it natural to say a thought is lingual at all, or, more strangely, is necessarily lingual. The intrusion of the word *proposition* is itself strange. The written language of theorists may be largely propositional but to insist on this as the form of thought in general attempts to straightjacket something that struggles in confinement. The word proposition already prejudges a lingual outcome.

To make a little progress with this requires a couple of things. **One** is being at one with thinking: knowing what it is by thinking. Perhaps this is like fishing. As soon as the brown trout is pulled from the stream it has an unreal clarity unknown in the vague swim of the river weeds, in its perfectly adapted to, watery medium. We all know this or can know this. The **other** is that thought is empirically given, something that takes place within what is physical. All of science has to be, in the end, dependent on our empirical judgement and we do have empirical access to thought in our physical, living selves. However, the **one** can mislead as to the **other**, unless we are careful. The problem concerns knowing thought and communicating thought. There is no direct transference of thought. Asked to convey thought we reach for words or begin drawing, or gesturing and many other ways (conventionally established mediations). And speaking loosely there is a temptation to say the words or the drawings or whatever represent the thought. But this response is a response to being asked to capture thought. The hesitation over *representation* presents when considering that *'I was thinking of the moon in the daytime sky'* (a verbal proposition) represents or is a representation of my thinking of the moon in a daytime sky. There is some relationship between the two but not that of the one representing the other. The one does not stand for the other, the one is not like the other. More appropriately the one **tells**, **states** what I was thinking. But how was the thinking itself? Do we not know that in order to be said to think of the moon in the daytime sky we do not have to perform some quasi-act of internal saying, or, more fantastically, mental writing? This is not to say that we cannot entertain such quasi-simulations. Nor do we have to manufacture some real image of a moon in the sky. We may seem to see the moon in the sky but there is no seeing and there is nothing to be seen. Instead what we have is a thought. We can think of the moon in the daytime sky without words, without pictures. We do think this without any of the above. We are approaching the crux of thought. A fundamental property of some matter. Looking back on what we have thought we know we were thinking of the moon in the daytime sky but we do not have to remember there being some internal saying (seeming to say) or even seeming to see, although both of these are distinct experiences which we can have. The entreaty is to look back and think. There will be so many things we were thinking of, all intertwined and coalescing, the thinking so complicated and often so incomplete that there could be no proposition or picture that would stand for it or even communicate it (what it was that we were thinking). Despite this we know what fragments there were; our access is empirical. And to broaden the notion of thought and thereby to clarify. Is there difficulty in ascribing thought to animals? I do not wish to answer this question here but if they have thoughts what is conspicuously missing from their repertoire is a symbolic mode. So it is possible they think thoughts, but they have no words nor pictures for what they think about? The dodge to neural symbolism would be conceptually slipshod. Clearly there is no empirical awareness for any thinker about any thought of any neural symbols. If there is animal thought is this then what we share? We can think all the things that we think without representations being the means by which we do it, just as we can see and be conscious of everything around us without representation being the means by which we do. Seeing and being conscious and thinking, as described here, are what happens in some material things. Of course we can say, more or less, what we are thinking or what we have been thinking about but the saying is not necessary for the thinking. We can think everything we can say in words without the words.

Words communicate thought but are not thought itself.

What though is the point in labouring this? The point is to bring into focus **thought** as a fundamental property of what's real and as basic as what is most basic, i.e. not something revealed by hidden layers beneath it. **Thought** is the stopping point. If we want to know about it we have to attend to the vast variety of its expressions. But naggingly do we not have to raise again what is influential but being opposed, do we not have to tell a neural story? No! Not at all! Well we can, there is such a story, but this story is not a reduction of thought to the core of what it is, instead thought is an enlargement of what's neural, what, under certain conditions, it amounts to. But a certain neural constellation brings thought into being and without it thought cannot be. Of course all that I am saying here is much too assertive and does not express what is a tentative struggle towards making what seems obvious and ordinary theoretically illuminating and radically alternative. For the moment there are two bothersome problems. One is the clarity of thought which is non-lingual and non-pictorial, about which and perhaps necessarily nothing can be said. The other is how there can be neural underpinning without a fully causal reality.

There are many things about the first problem ... too many. Consciousness is thinking. Tying one's shoelaces is thinking. Looking out for the postman is thinking. Remembering moments from ***Gone With the Wind*** is thinking. Writing these sentences now is thinking. And when doing nothing, as it would be said, still we are thinking. To be not thinking is not to be conscious, but sleeping is thinking (if this seems odd consider 'In my dream I thought of turning the corner but thought better of it'). To be dead is not to be thinking. Death is the end of consciousness (although mindlessly disputed). Access to thinking is part of what it is to be thinking. So there is no special appeal to **introspection**, although introspection is an aspect of access. But although we have access, interpretation is far from straightforward. To be thinking of the moon in a daytime sky is, in general, to be thinking of something visual, thinking of something one might see. It is tempting to say this thinking is seeming to see what one is thinking of (when there is no moon in the daytime sky to be seen). Clearly in order to think of it I do not need to seem to say the words or some of the words in *'the moon in a daytime sky'*, or even seem to see these words or some of them. At the same time when I seem to see whatever it is in order to think of whatever it is, there is nothing there to be seen, there is nothing I see. But access does not have to report seeming to speak the words nor seeming to see them nor seeming to see the moon and the sky in order to know one has had the thought of or is thinking of the moon in the daytime sky. This then is a strange thing. Attend though, retrospectively, to one's thinking. Is it not the case that, and this is how we speak, the mind flits from one thing to another and with thoughts having coalesced and having half-formed without any awareness of words forming or of ghostly propositions or of images or of other quasi-sensations? Yet we know we have been thinking of and all at the same time e.g. the window cleaner on the other side of the house, of one's partner sitting in the study having the money with which to pay him, of the pressing need to make progress with drawing attention to the reality of thought, of wanting a cup of tea and intending to make it at about 4.30 pm, of the difficulty in breathing because of one's cough and so on and on. All of this in a flash, in consciousness, but none of it represented or even seemingly represented. The reality of thought! What it is to think! A fundamental property of one's physical being. As soon as one begins to fixate on this thinking, making oneself conscious of what one is conscious, it is a step towards communicating thought, and words and images flood to the fore in a way drowning the thought, as though there was nothing there other than a lingual or pictorial or auditory etc., event of some kind. But thinking of the world and for that matter what is not of the world is prior to all that. And to be very assertive, it is what animals do, what prelinguistic children do, what early humans and Neanderthals did. Insisting on representation as the substance of thought turns animals and children and early intelligent species into behaviouristic machines and ultimately because all the distinctions are wrong finds that it must treat representation as neural representation and so treat all consciousness as concealing the universality of behaviouristic machines. What is lacking is being able to stare into the nature of thought and see it as a living physical property in its own right. Not being able to face the actuality of thought is to suppose it must have a medium in which to declare itself and which it is. But though there is a medium it is the invention of thought and not its fabric. Thought is part of the physical order or if not order just part of physicality. It is one way in which things interact or relate to each other. Another way is by means of causality. It is also a way in which things isolate themselves from other things. We might say there is a causal order and an intentional order and all part of the physical world, the nature of physicality. Conscious things both think effects and think causes and as such their existence cannot be just accidental contingency. Physics fails to understand the world because it subsumes everything under causal order or variations of the same. Things think and relate to each other through thought. This is something other than blind contingency. Of course if we try to turn thinking things into something other than things, into spiritual substance say, then we deserve all the pitiless derision and snobbery of a meritocratic physics. Physics and spiritualism are imperfect understanding, both unable to grasp ordinary experience. And that some physical things can isolate themselves from all other things explains how they lose contact with ordinary experience and fall prey to fantasies, like physics and spiritualism.

And so we have some purchase on the other problem. Perhaps what needs to be pointed to is **metamorphosis**, although in this context it may be no more than a metaphor. Many things become other things. Caterpillars become butterflies, tadpoles become frogs, a bulb becomes a daffodil, an internal combustion engine with wheels, petrol, ignition spark and engaged clutch becomes a moving vehicle. All astonishing transitions when seen for the first time. But all explicable within a causal order. Without prejudging and at the simplest level what is being said is that at a certain point things with specific neural constellations, and there may be many species of the same genus, become conscious things. This much must be causal and herein lies the possibility of artificial consciousness. But at this point begins an intentional order. This is what is *metamorphic* with reference to physical things. It might seem tempting here to work up the concept of autonomy in order to complete the metamorphosis of conscious things and so take them out of the causal order at the point they become conscious. And this is valid for some conscious things as is argued in detail elsewhere (**Invisible Cells & Vanishing Masses, Block 2, Cell 6**). But what of animals? Conscious things, but does being conscious put these things outside a causal order in some way. Is there not a strong case for them being just behaviouristic machines, no more, no less? Or if not machines then having existences fully explicable through causal chains? And if their existence is part of an intentional order then is there any distinction between it and a causal order?

Consider two cataclysmic events, probably fantasies. A lovebird on the wing is taken by a swooping falcon. A tsunami rushes forward to flood and engulf a 'paradise' hotel. These are different orders of reality. The tsunami has no notion of where it is heading, in no way has it singled out the 'paradise' hotel: it has no way of doing this. Words that spring to mind are terrifying, accidental, unintentional, blind fate. The unintended consequences of a purely causal order. Our commitment to intentional explanation may lead us to attribute explanations that are clearly incorrect, like 'the gods are angry', 'Gaia protests the pollution of the planet', nature's revenge'. Orders of events that are other than random, accidental or blind chance. It is not blind chance that the falcon takes the lovebird, as though they just bumped into each other because of separate trajectories they were on. The falcon has to see the lovebird and intend to take it. And of course respecting this distinction is the great achievement of Darwinian evolution, although often disrespected by those who proclaim to follow him. The natural order is devoid of intention at the evolutionary level. It is a random progression or movement. How everything interacts accidentally even though as our example shows this accidental interaction is mediated by intention. Falcons intend to take lovebirds but if as a result their numbers multiply and they themselves are not sought by predators then lovebirds run the risk of extinction and in due course if falcons cannot find alternative sources of food they too risk extinction. This is not an intentional outcome. The extinctions would be random causality. So there are reasons for supposing that consciousness opens onto alternatives to causal explanation. But this is not to face the brute problem of neural underpinning and conscious states.

Clearly neural events are continuous throughout consciousness and continuous with a living brain. A cessation of neural events is a cessation of consciousness. It is then tempting to say that neural events cause our conscious states (thoughts, sensations, feelings, desires, intentions etc.). This has been disputed. In the late 1970s Libet, Eccles and Popper thought science favoured the hypothesis that the mind was ahead of the brain in time. A simplistic causal argument that causes are antecedent to their effects. But then later research has been taken to show that the mind is behind the brain! So in simplistic causal terms affirming that neural events must cause conscious states. What though of the continuous nature of consciousness? Can scientific research have any plausibility if it isolates discrete, conscious events to correlate as post-neural events. Neural events are continuous but so are conscious events. One of the great literary themes of a not too distant era was '*stream of consciousness*'. An unceasing flow of consciousness, like a stream. How it is to be alive. If a scientific account requires the structure of consciousness to be understood as a neural event x followed by a desire y meaning that y is caused/determined by x then it has no grasp of consciousness. It may be as difficult to disentangle the conscious and the neural as it is to disentangle a flowing river's whirlpools from their mathematically formulated circulation of H2O molecules. Even if this was possible and it is thought that the whole is made up of discrete, digitally identifiable moments on both sides of the divide, it is just as true that conscious events are antecedent to neural events as it is that neural events are antecedent to conscious events. Thus to press on with such clumsy modelling and so to illustrate the difficulties, if neural event x (presumably a particular complexion of neurons firing in the brain) precedes the desire for a millefeuille, the desire for a millefeuille will precede neural event y which will in its turn precede intending to go to the patisserie which will precede neural event z which will precede rising from the seat, etc., etc. Where in this does one start and where finish? Which is the cause and which the effect? And of course the whole model is wrong. What kind of a mind could convince itself that this was correct? Well I suppose one that cannot comprehend that physical things are conscious, which in our ordinary lives none of us really has any difficulty in allowing (in saying good morning to our friends and destroying our enemies). We are physical things and we are conscious.

If this is presupposed in our dealings with living things what could we mean by it? What could it mean that physical things are conscious? Is this question at all adjacent to Nagel's questions about bats, not lovebirds, not falcons. **What is it like to be a bat? What is it like to be a bat, for a bat?** A bat is a physical thing. Is it conscious? Nagel's question make little sense if it is not.

We do not ask *What is it like to be a radio telescope?* or *What is it like to be a radio telescope for a radio telescope?* Except, perhaps, in absurdist fiction.

The **like** in all these questions is not strict. We are not asking usually for something to be named which is like, that is resembles what it is we are asking about. Instead we ask for a description, but a drawing in appropriate cases would do, and, of course, a drawing in some sense will be **like** what it is a drawing of.

Instead of asking what x is like we can ask what x is, and these questions in ordinary speech are almost equivalent, although in asking for unnatural precision we can grind out differences. And, of course, looking back to the lead question we can ask differently *What is a bat like? What is a bat?* And probably *What is a bat like for a bat?* (meaning something like *How do bats experience other bats?*).

The **Nagel** use of **bat** for a sample question must indicate that he thinks the question gains in significance if senses other than the five senses are involved (surely it can have nothing to do with a man who is known as Batman). It is a difficult case for which to frame anything that looks like an answer. We have no sensory experience of the bat's sensory world and yet there must be something that is the bat's experience. There must be something it is like to be a bat for the bat, but something we can have little grasp of because it is something we do not fully share.

Is the question any less difficult if it is *What is it like to be a human, for a human?* or *What is it like to be me, for me?* In all these questions the word that is central is not **like** but **be**.The question is then, perhaps, *What is it to have being or be a being, and if it has it or is a being what is it like?* And could a being not be conscious? Well yes when it is not conscious! But the question extended to *what is it like*, prima facie presupposes consciousness. So, it would seem, in asking what it is like to be a bat for a bat, we are saying that a bat is a being and is conscious and we are asking what that is like for the bat. Asking these questions and given a certain philosophical tradition preloads answers in the direction of describing, something assumed to exist, namely specific *qualia* in the bat's experience. As though it is this which we have to unravel and it is this we have to specify in all other cases pertaining to **the hard problem**, even if in the bat's case we have no means of being able to. This turns consciousness into a subjective content as opposed to a **contentless subjectivity, identical and indiscernible within a species**.

But if there is this alternative what does it mean to say that consciousness is a contentless subjectivity, identical and indiscernible within a species, and does this help with the other question about what it means to say physical things are conscious? And in what way are these questions made explicable by non-representational thought? And how does all of this alter the entire landscape of discourse about consciousness?

Contentless subjectivity is the minimal existence of consciousness stripped of personal identity. The possibility of which cannot be avoided. Empirically we have cases, devoid of memory and personal history but conscious. A bare individual. Perhaps a tabla rasa. The philosopher **Derek Parfit** said that what mattered in the grand scheme of things was psychological continuity, meaning things like continuity of memory and character, and not personal identity i.e. not whether DP was in fact DP. But there is something before all of this which is oneself and this is a conscious particular, a particular physical individual having consciousness. And this consciousness has no identity other than belonging to a particular physical individual. This is what we are before all else, conscious! A concept of bare consciousness. I suppose most amnesiacs (e.g. the lobotomized) retain something from before the amnesia, e.g language or a language, not just a language facility, but the opposite is possible and obviously enough babies lack language though conscious, or conscious for some of their time. And the concept of non-representational thought makes way for a bare but active consciousness. But could consciousness here be a red-herring? Should we consider that our life as a being, or the life of any being is somehow before or independent of our consciousness? That somehow we are continuously whirring around, like a computer when it is said to be idle? Thinking but not conscious? The notion of the unconscious, perhaps reinterpreted by computer science? Is there not testimony for this? The person bothered by a problem who is suddenly made aware of the solution, as though some unconscious thinking has been taking place which provides the answer? But as we begin to expand this perspective we have to be aware of the trap, the trap of expecting there to be some pedestrian explanation for what otherwise seem to be inexplicable creative leaps. The problem begins to be familiar, namely the inability to credit seeing, thinking, creating, as things in their own right, instead we are asked to reinterpret them in the guise of mechanical reductions. Yes there is dreaming but that is not a state totally devoid of consciousness, more it is a state where some forms of consciousness are not available to us. Yes our physical conscious state needs regularly to shut down in order that we can resume being physically conscious, but this is the explanation of periods of unconsciousness and not some life of unconscious thinking that solves all our problems, yet does not explain problem-solving, but instead, because it cannot explain it, puts it out of sight, sweeps it under a carpet and labels it mechanical. In fact shutting down is probably the wrong metaphor, resting probably is better. Thinking is dependent on the brain just as arm-wrestling is dependent on muscle and both of the things depend on or require periods of rest if thinking or arm-wrestling are to continue. This is a quite different order of explanation from computing processes, like *defragmentation*, also required for continued activity. Incidentally dreaming standardly is not experienced as any kind of organisation or reorganisation of files (or even memories) but much more as off the wall, creative leaps of imagination with only a tenuous relation to anything that might have been written on the tabula rasa.

A conscious, contentless subjectivity must fill with content necessarily but the content is contingently given. Consciousness is conscious of, conscious of something. And consciousness is some kind of continuity, no matter how fractured or fragmented but standardly neither of these. The continuity is of now and then. Consciousness exists remembering. Objects of consciousness are both now and then, directly given then held. The holding is variable but there must be some holding. The accumulation of content constructs an identity but this identity is contingent, but a necessary bare identity goes with a contentless subjectivity as well as a subjectivity with content. This bare identity is the consciousness of a physical particular, and its identity comes from the physical particularity. A consciousness as monad. A consciousness belongs to a physical particular and this is the only uniqueness it necessarily has. It is the uniqueness of a point of view and a having to live. This is what we, as beings, are, before all else. Within a species this is identical and indiscernible apart from the uniqueness of the physical particularity. The monads necessary point of view.

Surely this cannot be correct! Isn't such a view a debasement of humanity, a debasement of the rich individuality of persons, their personality, their character, their uniqueness? Clearly what any individual encounters in the world is contingent. So any individual could be picked up and dropped arbitrarily anywhere, like **Kaspar Hauser**, but how they would deal with this would vary depending on, depending on, depending on .... Well, something given? Something say innate? Perhaps something in the DNA? How would this fit with bare consciousness? And anyway think of basic needs. How does hunger enter consciousness? Does consciousness just find these inner compulsions like it finds the earth and the sky and the sun? Or are these compulsions somehow already part of consciousness and what does this mean for any concept of bare consciousness? But in raising these questions something essential is being lost sight of. It is not as though the idea of a physical particularity is just a physical location with some kind of camera attachment or some kind of sensory facility. The physical particularity is alive and conscious or a living consciousness. The physical substance of a person is conscious. And consciousness is not just recordings or a collection of representations but is thinking and remembering and feeling. All the time, from the very first moments. And in this way a content builds, but a content that can in exceptional circumstances be lost. This consciousness of a physical particularity comes in many different forms and will have a history of evolution buried in it. Just as in normal perception our gaze greets a given world, partly of our own making and partly not, so is our consciousness of hunger or pain or sexual desire or fear, although all of these are not just objects of consciousness but consciousness itself. This consciousness is not simply a recording of or a recognition of but is an active doing in ourselves and in the world. Consciousness is thinking, something that is constantly moving us on. And so decisions are made and things done of which we are conscious and so a particular life is constructed and commitments formed. But much of it could be lost and for the most part what we would be left with would be a particular bare consciousness: still ourself. At any particular point what we are is what we have worked up and worked out (because conscious life has to be ongoing) when always we could have done differently, although this is not to say we would have. The driving force for all this is identical from subjectivity to subjectivity apart from what is specific to its physical location, but this difference is immense. Each consciousness is different, but only contingently, because of what it is conscious of and what it has done with this. If all of this is correct it should be obvious that there is no room for anything that might be called a soul.

Thought is then a real property of what exists. As real as anything else. And thought is conscious and consciousness is thinking. Because our standard way of expressing thought is by means of language we are conditioned to think of thought as being a representation, but a few moments reflection should convince us that this is not so. We are trapped by the imperatives of expressing and communicating where automatically we reach for propositional forms. We have devised means for the expression of thought and have forgotten how the thought is that which is with us that we go on to express. But humans are just one variation of thought and consciousness. Nearly all of us will be familiar with natural history documentaries, they are hard to avoid. And there we see non-linguistic beings conscious and so thinking. We witness behaviour closely resembling human behaviour. Fashioning tools, solving problems etc. Things achieved by means of thought, thought as a property of the world, and the thought non-representational. But we do not have to focus on what we term higher-order skills (the word '*intelligence*' invariably enters the discourse). Let us revert to the concept of bare consciousness. Let us make the bold claim that bats are conscious and so think, and so, for that matter, are and do ants. Apart from the way some of science might wish to explain the perception of ants and bats what thinking takes place, if thinking does, can hardly be representational. But as we stretch to find a different way of talking about the subject the sceptical intervention will take the form of reducing the behaviour of ants and bats to the functioning of machines. We will be able to simulate what an ant does with a machine, perhaps using what we have discovered about ants to make the machine. So in finding directions ants navigate in some ways by the sun, and this could, in some way, be incorporated into a machine to achieve the same effect. And at this level such an account sounds plausible. An ant is, we discover, just a machine. But this won't do. What we have failed to rule out or consider is that the ant does what it does by means of conscious thought ('conscious' virtually redundant here), that it achieves what a machine might achieve but in a way that is operationally different. There is the possibility of there being different ways of following the sun. Perhaps a difference similar to getting to Oz by a sat-nav guided robot and getting there by looking up at the sun (or a rainbow). Surely though this again is something preposterous. How can one say an ant both is conscious and thinks? Is it not clear that an ant like a machine has drives that make it hell-bent on doing what it does, just like a machine? How carefully though do we observe ants? They are not consistently hell-bent, they do pause, perhaps look around, perhaps experience moments of bare consciousness. Of course, all of this could be simulated and a machine might require static moments waiting for its programme to take effect before moving off. However ants and machines enter the world in very different ways. Ants are part of the evolutionary world to which we belong, and this world is a continuum full of thought and consciousness and why should ants and bats not share in this. But then the argument in some science presses to ask how does the physical world become conscious. No scientific account of this is available or even theoretically possible. It must be that consciousness is some kind of fiction, some part of 'folk psychology' that we should have superceded. It is at this point that my account finds itself illuminated by the wisdom of peasants, and becomes what is, to my mind, truly astounding, and something I could in no way have anticipated at the outset. MAGIC also is a property of the world! A physical principle! **Magical Materialism**. But this will need some explaining!

'Closely connected with the peasant's recognition, as a survivor, of scarcity is his recognition of man's relative ignorance. He may admire knowledge and the fruits of knowledge but he never supposes that the advance of knowledge reduces the extent of the unknown. This non-antagonistic relation between the unknown and knowing explains why some of his knowledge is accommodated in what, from the outside, is defined as superstition and magic. Nothing in his experience encourages him to believe in final causes, precisely because his experience is so wide. The unknown can only be eliminated within the limits of a laboratory experiment. Those limits seem to him to be naïve.' **(John Berger, Pig Earth, Kindle, location 126-133.** My underlining.)

A number of thoughts need to coalesce here. To say it outright before explanation. Non-representational thought is itself a magical notion yet part of the material order. We think of whatever it is we think without anything standing for what we think as we think, although what we think we can express by means of something, something like language, or for that matter by means of language. And consciousness is just suddenly there not caused by a material state but suddenly a property of a material state, as if by magic, or by magic. And, in thought we can suddenly move on, know how to go on, go to something we have not thought before and maybe something that has not been thought before, as if by magic, and this, something way beyond anything in the empiricist theory of creativity. To say 'by magic' is to say by *means of the unknown* (which is only metaphorical), and if by magic it is unknowable or a brute fact. *The advance of knowledge never reduces the extent of the unknown.* A principle of material existence. Perhaps some scientific theory loosely gestures towards such a principle although unlikely to recognise this translation into ordinary language and ordinary experience.

The idea of magic straddles *the in principle unknowable* and *trickery*. Peasant and primitive belief in magic is an ongoing, active tradition. Within advanced societies there may be, even with sophisticates, a hankering for the existence of magic and the unknowable, but what prevails consigns magic to delusions of the ignorant. For the most part magic within the advanced world is accepted as a skill of professional tricksters, those who make it seem that something unknowable has happened but where everyone knows this only appears to be the case and in reality is the result of some ingenious sleight of hand, and therefore open to rational, scientific explanation. However, what I introduce here, although positioning itself alongside primitive belief, is fundamentally a conceptual dimension. What there is, is one way and not another way, even if the way it is contains a multiplicity, i.e. one multiplicity rather than another. There are brute facts. This is magic. Primitive thought is not without scientific inclination. To invoke intentionality seeks rational explanation. So what there is, is to be explained by the actions of gods/spirits, but magic enters this thinking when it is supposed that simply by intending to part the sea, the sea is parted. It is not known how an intention to part the sea or to bring about an earthquake or erupt a volcano achieves its effect. This is magic. What is fundamental to this concept of magic is that something is the case and it is in principle unknowable how it came about, even if it is accepted that the will of Zeus is an explanation of it. If Zeus had a laboratory and was experimenting with electricity and so was able to light up the sky with lightning having constructed a generator then he would be a boffin not a god. At some point the nature of existence is one way and not another or just is the way it is. At some point there are these brute facts. These facts are the starting point of scientific investigation. If you like these are empirical facts. The way things are. What exists has to be a certain way, it cannot be without its properties. It is its properties. Is this like what Wittgenstein asserts, '***The world is everything that is the case***' and '***...whereof one cannot speak thereof one must be silent***'? The nature of existence is the way it is and at some level this cannot be explained and if this is magic it compels us to be silent: rational and scientific thought is no help with this. What is magical is that things have to be a certain way, and to say 'magical', as intended here, is having to accept that things, at a fundamental level, have to be one way or another and that is **why they are**.

But this is a hopeless position if taken to mean that because everything is the way it is and not another way then magic explains (no explanation at all really) everything. A property of being a giraffe is having a long neck. Magic is not the explanation of this as it would be in a creationist narrative. An evolutionary story explains the long neck. This is not the same as saying where as a result of an evolutionary process or a causal process a property is manifested then evolution or causality explains the property. What is needed here is a distinction between necessary and sufficient conditions. It is natural selection which adds to evolutionary and causal process sufficient conditionality in the case of the giraffe. Consciousness has to emerge in an evolutionary and causal nexus but neither of these explain the leap of consciousness. They are necessary but insufficient for consciousness. This suggests some fundamental distinction within what we call properties. Whether *property* is the best term for what is being discussed will be faced in the next section, but for now it makes for general intelligibility. Here I am assuming (and this has to be returned to) that a world containing a variety of substances would be differentiated scientifically and so fundamentally by means of atomic distinctions. I have used the term *brute facts*. These would be the brute facts of what exists. So copper has a specific atomic structure which differentiates it from other substances and this is a brute fact about existence which we just encounter. Lots of copper in the world, in the universe, but all of it differentiated from say iron by fundamental properties which copper just has and has to have to be a differentiated substance. And how could it not be a differentiated substance? This, magic. This, the unknown. The, in principle, unknown. Consciousness needs to be conceived of as such a property. A fundamental property of evolving, material substance. As a fundamental property there is no reason why it should not have its own properties just as atoms have protons, electrons, neutrons, neutrinos, quarks etc., one of which for consciousness would be non-representational thought, the subject of this section. And if there appears to be a category difference between subatomic particles and non-representational thought or direct perception so that there might seem to be a difficulty in equating them as fundamental properties it is the case that the atomic and subatomic have a presence in primary and secondary qualities. And the difficulties this raises with regard to identity, causality and other possible relations is for a later discussion.

But do recent advances in, as it is alleged, communicating with those suffering from *locked in syndrome* raise difficulties for a concept of non-representational thought and as a consequence for some of the issues of identity and causality just referred to? Those with locked in syndrome exhibit no behaviour but are alive. The question arises whether those with the condition could be sensing as well as being physically alive. There has been no way of telling. However monitoring brain activity has led to the claim that sensing does take place, and that thoughts about what is sensed can be read from monitored brain events. The present level of monitoring allows for limited results but nonetheless, if true, very important results for those afflicted and those who care for them: the possibility of establishing a line of communication. Those with the condition can be asked questions, and, it is claimed, brain events can be read indicating positive or negative answers to these questions. So a patient can be asked whether or not they like chocolate. It will be known from preferences known about before the onset of the condition what the answer should be. Presumably correlations between brain events and negative and affirmative corroborations have been established with normal subjects. The proof of consciousness is taken to be established by the occurrence of the pre-categorised brain events (say affirmative) and the previously confirmed preference for chocolate. Everyone is held to be happy with this outcome. To simplify, it may have been established with normal, conscious subjects that a particular area of the brain 'lights up' when a subject responds 'yes' to **yes** or **no** questions, and another area of the brain when the response is negative. Of course the locked in subject, if conscious, may have a change of mind about chocolate following the other changes that have occurred, or it may be that a particular brain event can occur (i.e. is logically possible) without any attendant consciousness. The normal and conclusive methods of corroboration are not available with a patient suffering locked-in syndrome. But what is claimed seems probable, especially if there is invariable correlation over a range of questions. I am assuming that the detected patterns of brain events are not language specific and so apply universally across a range of languages. On this interpretation what the brain event establishes is the occurrence of affirmation or disaffirmation. The fact that actual minds are invariably non-binary when faced with black and white questions is a complication any science of this area would need to address.

However, turning back to the subject on hand, if the science is pointing to ways of detecting affirmation and disaffirmation across many languages then this would seem to be evidence in favour of thought being non-representational. We might then be able to affirm or disaffirm regardless of whether or not we possess a language. And this possibility seems, when one thinks about it, unquestionable. A difficulty arises though when thinking about how to interpret the relation between brain event and thought. Simplistically we might be tempted not to examine the depths of this relationship and to suppose that the relationship must be causal. If this was so, conjectures about the magical nature of thought might seem to have no foundations whatsoever and this would not come as a surprise to most in the educated mainstream. The cause of affirmation (thought) would be a specific brain event (area of the brain lighting up). But as previous discussion has made clear we have alternatives to this. The relationship is like drawing a dancer and the bare, physical movements necessary and possibly programmable to a computer system to achieve (in a computer system more or less) the same physical outcome. The physical movements in programmable language are not the cause of the artist's drawing, and the artist drawing is itself a physical process in the world, the identity of both are shaped by each other. Different but related properties of the same thing. Affirmation is a physical event shaping the identity of brain events, which in their own right shape the identity of affirmative events. Causality is too crude a model and leads too easily to deterministic rather than magical scenarios. Although that a cause has a particular effect has been held to be, even within scientific traditions, a correlation fundamentally inexplicable. Many of the misunderstanding and complexities of this are addressed in **Invisible Cells And Vanishing Masses**.

Non-representational thought then is one aspect of materiality coming to life, being alive, living. As if by magic or by magic. Not requiring a pre-programmed language. Aladdin rubs the lamp and the genie materialises. Of course this does not happen except in the story but the story contains an essential and undistorted condition of real magic (if any such thing is real). Aladdin rubs the lamp for the first time without intention, and the genie appears, much to his astonishment, and there is no explanation of this, it is just a property of the lamp. Things can have just the properties they have because to be the things they are they have properties. To approach the world scientifically is to be predisposed to thinking that everything that there is can be explained despite the innocent paradox of infinite interrogatives. As an attitude this is a fine approach to the world, it has obvious pragmatic benefits, but we should also consider that things may just happen, just as they may not happen: the fundamentals of ***being and nothingness***. There is a new born baby and it is alive, and full of thought, from the off, not a representation in sight. So there are things where we can explain why they are so and there are things that are just so: it is where explanation runs into magical materialism. As a result that which just so happens to be so, is not shrouded in some spiritual mystery, it is always empirical and sensible. And this is so because it is the material world.

IRREDUCIBLE CONSCIOUSNESS

The starting point is that consciousness is an undeniable, empirical datum, as much part of the world as anything else we consider to be part of the world. We encounter many things in the world and one of the things we encounter is/are conscious beings. It does not need to be argued that our world is a physicality (something open to expositions in hard science, but perfectly intelligible without it). Within this physicality are physical things that are conscious, undeniably, surely. Or we could say, with no real difference in meaning intended, that the material world contains material things that are conscious, undeniably, surely. What is intended is that to say such things is fundamentally different from saying there are two orders of reality, physicality and spirituality. Towards the end of Honderich's **Actual Consciousness** he brings to the fore '*the hard problem*' or the mind-body or mind-brain problem. He does this to point the way towards distinguishing between an approach to consciousness which focuses on correlations between the neural and the non-physical and a thorough going physicalist approach. Being a physicalist, **Honderich** argues that it is necessary to get rid of correlations between the neural and the non-physical, the reason being not that there are no correlations but that the final term of this correlation is missing. This, **Honderich** claims to be a position shared with **Searle, Dennett** and **Papineau**, and so a position not without authority. However, for

Honderich the missing term does not leave us without correlations, or leave us with only neural mapping. Instead he posits a subjective physicalism leading to a distinction within physicality, a physicality he claims to contain *'real difference'*. The particularly strong aspect of this is the invitation to readdress physicalism, or to get a clearer understanding of what is physical. The weak aspects of the argument from the perspective of a readdressed physicalism are the subjective reality of perceptual consciousness and the representationalism of cognitive and affective consciousness, both of which are held as real differences within physicality (physical but different). The problems with this will surface later in the commentary on **Honderich'**s book. For now it is sufficient to indicate that the subjectivism and representationalism as concepts integral to Honderich's actualism seriously hinder any breaking free from the philosophical traditions that have constantly muddied and muddled our understanding of consciousness.

Adjacent to **Honderich**'s concluding remarks on physicalism is a brief consideration of **pessimism** and **mysterianism**. The problem is how consciousness is generated by the brain. Eminently, Nagel, Chomsky and Mcginn suggest the problem is insoluble, that it remains and will remain a mystery, something beyond rational discussion. For a moment this seems to chime with the considerations of my previous section, i.e. *non-representational thought,* where **magic** surprised the discussion. But neither pessimism nor mysterianism are at all implied by this intrusion of magic: instead it is offered as being both rational and useful. There is a phrase used by Honderich amidst these closing remarks which is helpful, namely **'Put up with reality as it is**'; not that, with reference to this problem, he is inclined to. It is **Honderich'**s conviction that science will be able to explain how it is that consciousness comes about. Maybe his attachment to determinism (which he may equate with rational, scientific explanation) makes this outcome inevitable for him. What I am arguing is that the explanation already stares us in the face. Consciousness is an irreducible property of physical reality and it is this we have to *put up with* it. We are not waiting on scientific discovery for an explanation. Clearly we may get to a position of much more detail than we currently possess about the neural correlates accompanying our streams of consciousness, but that will not be an explanation of those streams or of how they come about: all we will have will be accompaniment. Although as properties are added so the reality changes. Is this like tears accompanying being upset or being upset accompanying tears, where tears do not explain being upset nor being upset explains tears, but both are properties (is this the best word?) of a physical particular and are the accompanying realities that they are? Clearly being upset changes in some way the nature of tears, e.g. not the tears caused by a fierce wind, not the tears secreted and shed by actors, etc.

What is required here is a common sense account of physicality, which is neither deep physics nor abstract metaphysics. The physical is many ordinary and extraordinary things. Solid objects, liquids, gases, flames (the archaic, but accurate, **earth, air, fire and water**), heat, odours, tangibles, light, gravity, radio waves, radiation, magnetic fields, plasma, molecules, compounds, elements, atoms, electrons, protons, neutrons, hadrons, quarks, leptons, bosons, gluons, photons, movement, stillness, growth ... the list goes on and on. Another list could be of all the physical particulars in the universe, starting with all the things on my desk, which in itself would require, at the very least, computable powers, although the starting point would be simple, e.g. Chromebook, all in one pc, mice, mice mats, post-its, pens, notebooks, keyboard, table lamp, sewing-machine, cables, duster-brush, A4 paper, Kindle, desktop and so on. The initial list is a relatively arbitrary collection of generalised physicalities. The second list a collection narrowed initially by specific spatial location. The first list is a mixture, mainly generalised collections of ordinary physical particulars and generalised collections of particulars which scientific investigation has uncovered. This distinction is also, in part, a distinction between compounds and what compounds are composed of (perhaps suggesting that an investigation of **Leibniz**'s **Monadology** might prove relevant, and, perhaps, tempting us into some sloppy thinking about what real, physical things are). Is there anything general that can be said of all the items that might appear on such lists? Something that is true of anything that belongs to the lists? Space and time come to mind. Not in any abstract sense, but in the sense of now and then and here and there. Can we say with certainty that all of physicality has a position in space and time (as specified)? Time we might say is necessary but is hardly sufficient as so-called spiritual realities have to be thought of as occupying positions in time: being eternal (at all times) being an alleged property of many alleged spiritual realities. Spiritual realities also, it is alleged, have spatial locations attributed to them and can take on physical forms. However the spiritual, it is alleged, is not necessarily spatial, it has a possibility of some other mode of being, mode of existing. Physicality, on the other hand, is necessarily spatial and temporal. So is this the all encompassing duality, the physical and the spiritual? One term of which, I will hold, does not refer to any reality whatsoever, and if this is so does this just leave us with physicality? Is it that all that exists is physical! Is it that there is space and time and the things existing in space and time (which have to be therefore spatial and temporal things) and there can be nothing else? Is it that there is just **the physical and its properties**?

The incoherence of the spiritual becomes apparent as soon as conjoined with the spatial. Jesus Christ, if he existed at all, was a physical body with temporal and spatial locations, but, presumably, resurrected as pure spirit, although re-appearing as a body but not being one prior to ascension. Mary the virgin is visited at a particular place and time by an angel, who, presumably, is not a physical body but looks like one, and then she is visited by, as the angel announced, the holy spirit, not a physical body, who impregnates her somehow or other with a god-child. This mixing of the spiritual and the physical is then within Christianity extended to all human beings in so far as they are credited with being bodies containing spirits. However the spirits that they are, are not properties of the bodies that they are because the spirits are destined for non-corporeal existence (the spirit leaves the body). All this spiritual existence within the Christian tradition, and for that matter in most of the world's religious traditions, then enters and assembles (both concepts we only really understand through physicality) in some spiritual realm (again primarily a physical concept). In primitive thought the spiritual realm is conceived of as a sort of physical place somewhere beyond the physical place in which physical bodies reside. Medieval representations of the totality of existence portray some arrangement of spheres spatially related to each other and thus conferring by default a kind of physicality on spiritual spheres. Sophisticated theology realises all too well the dangers of such a topology given that scientific investigations reveal no traces of any such spheres or locations. The notion of a quasi physical place filled with millions of quasi bodies, supposedly moving around, becomes an unmanageable narrative. Instead theology resorts to a metaphysics of abstract types and tokens, an existence of pure forms supposedly made intelligible by a very abstract *form* of love. Such an account only seems plausible when set against the total implausibility of quasi physical spheres etc. And *form* itself is primarily a concept of physicality.

But for the purposes of addressing consciousness the question of **all there is** does not have to be oppressive. It is not necessary to fully work out the limits of physicality. Clearly there is physicality, and probably not confined to materiality, and clearly not spirituality. The challenge of idealism has passed and a long time ago, and when it was it was linked to spirituality, which is an incoherence without metaphorical use of space and time. A hopeless alternative. What exists is physicality and not much more, probably nothing more. There are physical things. Things which at a certain level of description have chemical and atomic structures. Things which occupy space and time. These things have properties also referred to as qualities. We could say characteristics or attributes. We have to be careful though because physical things are properties (clustered) and nothing else. To think otherwise is to open metaphysical vistas, peering to construct underlying substances (we know not what) displaying properties we do know. To take a simple example, hydrogen. Hydrogen is a chemical element with a standard atomic weight of circa 1.008, the lightest element on the periodic table, colourless, odorless, tasteless etc. There are innumerable other properties including those causing quantum effects like the wide spacing of rotational energy levels because of hydrogen's low mass, or signifying its low mass. This is hydrogen being hydrogen. How we find it. The properties which make it what it is. The physicality of hydrogen. Its existence in space and time. Physical things are the things which they are in virtue of their physical properties. What is being denied here is a sort of reductio ad absurdum. Whereby there is no stopping point. Where everything reduces to something else, but, of course, impossibly. The denial of this is the assertion of irreducible properties. Hydrogen's low mass is an irreducible property, part of what a hydrogen element is. Why a hydrogen element is a hydrogen element. But in order to illustrate itself the argument already has reduced itself to the level of simple elements. The point being made is true of this level of reality but is also true of compounds, even complex compounds. The properties of individual things, simple or complex.

There are problems here though. Problems in specifying what the argument argues. One issue is that of reducibility and magicality. Take a wall, say a brick wall, or even a simple elevation constructed from toy bricks. Such a wall reduces to the sum of its bricks. Physically it is no more than the bricks assembled. Suppose though at a certain accumulation of bricks the wall entered a state of nuclear fusion, in this case a result of increase in mass. The reduction would no longer be a full deconstruction of the now fissile wall. Very simply the sum would be more than the sum of its parts. The sum would have physical properties that would not reduce to its components, although the components taken together spontaneously generate these properties. We might say this is a matter of cause and effect but the effect we get is just what is conjoined with the cause. This is what is magical. That things have to be one way and not another. That is how they are. This is how we find them and it would be dense not to accept the this sidedness of things. The extra dimension with consciousness, is that consciousness as an irreducible, physical property is no mere effect. It transforms that out of which it is constituted. Consciousness frees matter. And as a matter of fact that is just how things are, or some things.

Clearly consciousness is physical, empirical, just part of the physical world. Quite apart from ourselves we bump into it every day. It is all about us. In a world of things some things are conscious, irrefutably so, undeniably so. And the physicality of conscious things is constantly demonstrated by their own effects, the changes they bring to the physical world. Part of the physical, transforming it in ways that are different from the other principles of transformation that occur physically. Matter is the way it is with the irreducible properties which it has, but some matter is free to go in any direction and other matter is not. Consciousness is a property of physicality that makes this possible. Not necessarily all at once or in any form but through its development, its evolution. Much of science denies this insisting on a reduction of consciousness to a deterministic order of neural events. To resist this the present account draws attention to irreducible physical properties. But consciousness like anything else is a spectrum, individually instantiated. It is necessary for this account to be nominalist and not essentialist. The traditional and embedded presuppositions of essentialist thought creep into thought and understanding unawares and where this happens in this account it is necessary to impose a nominalist translation or filter and apologies are offered for any proofing that is insufficiently thorough, but embedded reflexes infect proofing too. Moreover, if the thought above of freeing matter seems underdeveloped this does not require further addressing here as its parameters are explored and its coherence demonstrated elsewhere in **Block 2 Cell 6** of my **Invisible Cells & Vanishing Masses**.

It is now time to turn to some of the sources for this collection of thoughts and to some of the texts against which the thesis advanced needs to be tested. Before this though I add a sort of postscript fleshing out to some extent undertakings littered throughout the text so far.

POSTSCRIPTS TO EXPOSITION

Concepts of mutability, anarchy, chaos, randomness, unpredictability, freedom have affinity, and, of course, science has to recognise the phenomena, or that there are events so conceptualised. The flow of water or liquidity in general is an appropriate image for the volatile and mutable and there is a science which addresses actual flows, namely Flume Theory. The theory though, to its credit, freely admits the disparity between mathematical models applying to laboratory conditions and field conditions (the actual flow of liquids in natural circumstances). The theory is useful in the management of actual water, especially where the containment of water is man-made, and even in purely natural circumstances it approximates usefully to actual flows, but there is an unbridgeable gap, probably quite small, between mathematical prediction and actual, natural flows, and even flows in man-made containment when there is a large scaling up from laboratory conditions. This echoes themes already explored about the generality of scientific theory and the specificity of actuality. It is to these issues that Chaos Theory is directed. It attempts mathematical proofs of the inherent unpredictability of events because of possibly infinitesimal variations in initial conditions. However, the theory, despite what it allows, does not depart from the basic scientific axiom of determinism. Everything is causally determined but universal predictability does not follow from this.

Events are in principle unpredictable but still determined. For the purposes of the discussion of this book all of this can be passed over without comment, until the point at which the theory's tentacles gain a grasp on conscious events. The unpredictability of consciousness is not now an argument against determinism. But the argument for autonomism is not based on unpredictability although unpredictability follows from its truth. The argument for autonomism is that in conscious events the unpredictability is explained by consciousness making matter free. The ultimate explanation for a conscious event is that it is freely chosen and not explained by some remote concatenation of variable, initial conditions. The proof of this for conscious material things is absolutely empirical. The fact that, if I want, why it is that I do what I do is just exercising my freedom of choice and the same applying to my wanting to exercise this freedom. It is the lack of compulsion and so the absence of determinism. And this applies to another threat to autonomism, namely Games Theory. Games Theory boxes its participants into restricted psychologies and then sets them playing games as though they were chess computers or dehumanised mathematicians. It cannot allow full autonomous conscious beings to enter the games, beings who might just refuse to play the game. In which case it is never describing conscious matter. Its participants are computing machines and they are predictable and determined. The absurdity of Games Theory as a description of autonomous beings is immediately manifest in its paradigm the Prisoner's Dilemma, where a few contingencies are subtracted, like discounting retribution from the gang, and a single driving motivation is left in place, namely getting out of prison asap, and then with this absurdly simple scenario the game has to be played. Two computers would have no difficulty conforming. Two conscious beings could not be so logically shackled unless freely consenting to be so, as do two chess players when they sit down to play a game of chess. The point about autonomism is that outcomes, x or y, can be fully explained by a decision: a decision that could go either way, the real possibility of being able to go either way without any reason other than exercising one's freedom. Determinism cannot allow any sense to this despite it being empirically manifest and self-evident.

So it is being argued that consciousness brings a non-causal, non-deterministic existence to physicality. But if consciousness is on a spectrum does the argument apply to all sectors of that spectrum? Empirical sense distinguishes conscious life from other life, although there are those (pantheists, Gaiaists) who would envisage a larger spectrum. There are though central cases where conscious life is indubitable, as certain as a universe of physical things and itself physically manifest. However, as covered in some depth in **IC&VM** the content of consciousness, what any conscious thing is conscious of, in the sense of what it is like in any consciousness, is private and necessarily so. The only certainty as to what it is like comes from the corroboration of the ones who are conscious. This privacy is an elemental part and an elaboration of consciousness being a magical, physical property. The content of consciousness is accessible only through consciousness, and part of the reason for this is that this content is non-representational thought and for this reason inaccessible to observation. For example, dogs are conscious. The denial of this would demonstrate the absurdity of some scientific theory. Moreover, saying that dogs are conscious is saying that they think, that they have thoughts. Computers do not think, unless they are conscious, even though they run representations, and so do not have thoughts, and when it is said, as it often is, that they do, what is said is metaphorical, the appropriateness of which being derived from their simulations or imitations. Dogs, however, must think all sorts of things, about food, walks, hunting, owners, strangers, other dogs, smells, drinks, barking etc., etc. Clearly dogs are aware of such things and to be aware of x is to have the thought of x. But dogs do not employ language or pictorial representations, even though they understand some human language, so their thinking must be non-representational, though potentially representational. It is possible to think of x without having any means to represent it. If this is so the next questions concerns whether dogs are part of a free physicality. And this must depend upon whether they belong to that part of the spectrum where there is consciousness of consciousness. The most unusual but clearest case of a free consciousness is that where x is chosen despite all the reasons an agent has for not doing x but where x is still chosen because it can be. To be in this position requires being conscious of one's consciousness of x or not x, and if this is present it will be a precondition of freedom accompanying all our thoughts. So what of dogs? It follows from the necessary privacy of consciousness that we cannot know. Certainly the training regimes imposed on dogs attempts to robotise them, but we know, in fact it is empirically certain, that dogs are, despite our best endeavours to eradicate this, disobedient. Determinism for dogs will insist that disobedience is caused by uncontrollable instinct. Consciousness interpreted as a magical, physical property of material things disputes this, but because of the privacy precondition means that it is something about which we must remain silent, certainly it is something about which we cannot speak, i.e. about the content of this consciousness, unless told and, of course, in the case of dogs and other similar conscious beings we cannot be. A free physicality in the animal world might seem disputed by the observations in natural history where often the story of animal behaviour is presented as one of endless repetition apart from evolutionary changes, and therefore inclined to be robotic, after all, in contrast, the hallmark of human history might seem to be progressive civilisation. However, much natural history is not long-term and where there have been long-term studies many unpredictable developments have been observed in particular groups of animals, like the Goodall surprise of finding chimpanzees starting to form gangs to hunt orana tangs, and the prevalence and development of tool making in many species. Aspects of creativity though may not be the point. More to the point may be accounting for why an otter, say, moves to the left and not the right for each and every occasion on which it does. What of wasps though or bats? Are they conscious? If so must they have thoughts? Could we entertain the proposition that they have thoughts about thoughts? These issues cannot be settled philosophically. They are empirical matters but this is not to say they can be settled by scientific observation. Some physical properties may be impenetrable. Perhaps dark matter. Very probably consciousness. Perhaps it is the very matter of what we are dealing with that cannot be penetrated, that the logic of what we are dealing with locks us out. So not everything is knowable for disinterested observers: from this viewpoint somethings are in principle unknowable.

Consciousness of consciousness, self-consciousness makes it possible for us to make everything about ourselves the object of our thinking, our thoughts. It is the foundation of autonomy. This way it is possible to escape the lasso of determinism, so that when everything we do is formally consistent with a chain of causes it is always against a background of self-consciousness and so the possibility of escape and indetermination. What is it to be conscious without consciousness of it? Is this to be back in a determinist order?There may be a number of concepts needed to be sorted, brought under control and related to each other in order to approach the problem. These include the concepts of soft determinism, the concepts of the unconscious and subconscious and the concepts of physical process.

To begin with try to think through the implications of thought being a physical dimension, just part of the physical world in its own right. The arguments of this book have tried to clarify and reveal these implications. Proceeding along these lines it is possible to conceive of spontaneously occurring thoughts and thoughtful actions leading to each other and explaining each other: a physical process as appropriate to animal behaviour as to human behaviour. Without self-consciousness it is possible to conceive of these processes being free in a soft determinist sense. The thoughts and thoughtful actions arise spontaneously (magically) and tie in with other spontaneously arising thoughts and thoughtful actions: a self-enclosed physical order, part of what happens in the physical world one that drives on the things so conscious. Driven from one stage to the next but not from this straightforward exposition constrained or compelled by external causation (external to the physical process of thought), although this story needs to accommodate the interventions of external causation. A passage from **John Berger**'s Lilac and Flag is instructive.

'I saw the marten this morning. He was running. He never runs blindly. He considers each mound in the garden before he jumps over it or skirts around it. When he skirts around, he keeps very low to the ground. Pointed, slim and the colour of flame. As cunning as he's quick. It was three months since I'd seen my marten. Where he lives I don't know, but it can't be far from the house. We live side by side but invisible one to the other. When our paths do cross, it's the result of an accident or a mistake. This morning he was being pursued by a dog.'

To ordinary understanding this is not an example of poetic licence or anthropomorphic. It is an account of a conscious, physical animal thinking its way through the physical world, free to do as it pleases, where 'as it pleases' comes from its spontaneously arising thought, but these thoughts confined to its particular, physical existence, and cut across by the external causation of accident and mistake. An animal driven but free, yet not fully autonomous. What this account is doing without is anything that attempts to instate a causal, interventionist process separate from streams of consciousness. Consciousness is a physical process of material things in its own right. So what is being put to one side are antecedents like the unconscious and the neural. Theory that requires antecedents cannot grasp the actuality of thought as an irreducible, physical property of physical things. In some ways the concept is one of accepting as a principle that there is nothing and then there is something, not being and nothingness, just common or garden nothing and something. And how unscientific is such a proposition? In the end doesn't science have to concede this, not about spiritual substance (a contradiction in terms anyway) but about physicality itself, the subject of its enquiry. The marten sees its prey and begins consciously to stalk it, thinking its way around the terrain, not because it is programmed (which in the straightforward sense it clearly is not) but because it thinks: there are different ways in which the same things happen in the world. Thoughts are integral to the physical condition of what has the thoughts, part of the expression of that physicality, not some unreal, quasi- real effect of some solid, physical cause. In what sense are unconscious neurons *unT1…n* firing, an understanding of stalking prey. What the marten is thinking we are precluded from knowing in **Nagel**'s sense of knowing what it is to think as or be like a marten. But why should it be far away from the way in which fishermen think when they strike because fish bite? In this latter case the thought is instantaneous, spontaneous, and it would be so clumsily conveyed by some precise, representational proposition, which is never formulated accept in retrospect, and even then involving draft modifications.

What is attempted here is to create some plausibility for a different way of talking and thinking about thought, consciousness and thoughtful action. I am aware it can be no more than a sketch and that I do not have the resources to provide a full treatment of all the problems and conceptual niceties of the subject. It has been hinted at that there is a fundamental distinction between intelligence and intellect and that intelligence might just be a machine-like function humans share with artificial intelligence machines, suggesting unconscious, causal processes culminating in solutions, problems solved, numerical answers. But if intellectual thought can be conceived of as a concatenation of somethings from nothings (the magical nature of consciousness) may it not be that intelligence in humans is just part of the same story? Is the idiot-savant the recipient of unconscious, causally-activated algorithms or the recipient of the magic of thought? And if an algorithm how does its completion in a human brain become a thought, and if the magical process of thought how do we explain differences between the gifted and the rest? And if intellect is something separate from intelligence might the explanation of this fit in with the way self-consciousness has intruded itself into this treatment of consciousness? Is intellect not calculation but standing aside from thought and reflecting on it? We attribute some intelligence to non-human forms of life but not intellect or intellectual process. A difference in autonomy? If it is thought there is any value in this approach it must be left to others to complete a full treatment of consciousness as an irreducible, physical property of material things, whereby these things have unmediated direct awareness and think non-representationally, moving magically and privately from one thought to another and in some instances finding themselves to be free material substance.

An unsystematic consideration of many of the perspectives this account opens onto will surface from my reflections on texts I have found relevant in producing this work and which are to follow. Although feeling a more systematic, more comprehensive and in some respect a more technical treatment is requisite, it should not be forgotten that we understand consciousness well enough without reflection, if only because we are privy to it. We are right to take it for granted. It is the danger of chimerical theory that creates mesmeric, **hard problems**.

Perhaps the best summary key to understanding what is offered here is accidents of identity. Individual, physical things have identity, partly determined by specifics of time and space and partly determined by properties in virtue of which things occupy time and space. If this was not so things would not exist.

What individuates a piece of copper from a piece of iron? We say they are made up of different elements, and this has entailments. But why is copper copper and iron iron? The answer must be the properties they have that individuate them. This is how we find them to be what they are, how we come across them, how they are there in the world to be come across. From the point of view of simply encountering something, knowing nothing of it apart from what confronts us, its properties are accidents, or they just are the properties it has, but these accidents are its identity. We come across something and simply accept it the way it is, simply accept its identity, the thing and its attributes. And, the argument has been, some things are things that think, **thinking things**. They are not other than things but some things think or are conscious. In a universe of things we can encounter things that think, where thinking is a property or attribute or quality of some things. In what sense is it accidental, in the sense that it is spontaneous. Things have to have spontaneous properties, there has to be something that it just is, and so in this sense and this sense only it is accidental, accidents of identity. I labour this point because a simple but fundamental shift in how we conceptualise is required to grasp this as a different way of thinking, and the more often it is said and in different ways the more chance there is that the flat earth will be seen as round, because it can be seen either way but it makes an incalculable difference which is true.

'*I am not an intellectual, I write with my body.*' **Clarice Lispector**. Forging a new shape, a movement, a shift in paradigms of theory is overcoming an immense resistance, even in those who see a need for it. There are those who seek a middle ground between materialism and dualism, not quite knowing how to establish it. But the appeal has been for a more liberal, nuanced concept of *mindedness*, as though this would crack the **hard problem**. And there are others who decry an anthropomorphic approach to thought in animals, but, in what they think is a more liberal understanding, ask us to be much smarter in investigating the smartness of animals (the notion that they might have different ways of being smart from which prejudice blinds us). But all of this is the problem. The problem is not one of mind and body, although that has been taken to be the problem in philosophy for centuries, the problem is that of bodies and bodies which are conscious. The starting point has to be the conscious body. Our bodies are conscious. And the brain is part of this interconnected materiality. Dualism is wrong and materialism is right but as it has been formulated limited in scope. If we reflect just for a moment and attend empirically to how we are, we are a mass of body consciousness, we might say, before all else. This has nothing to do with smartness, problem solving, intelligence, the mind, it is our consciousness of ourselves as bodily things. We are things that are conscious and this may be no more than a fragment, although typically it is much more.

HONDERICH: ACTUAL CONSCIOUSNESS

(What follows is a long commentary on **Honderich**'s **Actual Consciousness**. It is included because of what it reveals about consciousness. A Consciousness dealing with **Honderich**'s arguments, which arguments have similarities with those of my **Being Conscious**. But it is not a matter of finding something, instead it is seeing something for the first time, as if by magic).

At the outset one is not sure how Honderich's book is to unfold. How could one be sure, it is a long book. He seems to favour the word 'inquiry' (**Hume!**). Looking to the proposed inquiry naively, why '*actual* consciousness' instead of 'consciousness'. Standardly *actual* adds that the actual x is the real x, the very x, exactly the x, e.g. *I was standing at the actual spot where thirty years before I caught brown trout.* A hunch, though, as to what **Honderich** may be about, or perhaps it is what I am about. *The reality of consciousness is not what often is discussed as the subjectivity of experience, it is instead a property of the world and as real as any other property i.e. the world contains consciousness or even the world is conscious. Thus actual consciousness might seem to be a proper subject of scientific investigation. Put more strongly the claim would be that physical things are conscious or some of them are, or consciousness is physical. Thus consciousness is not something other than the physical universe, nor is it something that does not exist, something unnecessary to an understanding of reality -*

which is a contradiction in itself-. This is to have said much too little to have explained *actual consciousness* in this sense, but, if this is **Honderich**'s point then it is a dimension we need, and a dimension we need whether or not it is his point. In his introduction **Honderich** also distinguishes between a *figurative* account of consciousness and a *literal* account, suggesting that a figurative account roots us in common-sense but that we need to do better: we need to achieve a literal account. *Figurative* seems to be used as meaning metaphorical (perhaps then accounts of what consciousness *is like -and so linking to the various philosophical discussions about such things as what it is like to be a bat etc.-)* as though we have a natural commonplace language for discussing consciousness but that it is metaphorical and needs to be replaced by the introduction of scientific concepts, many of which we may not yet have developed. The suspicion is that this approach will tie in with **Honderich'**s long-term hankering for determinism and so deny any linking of consciousness to an account of freedom, or any linking of freedom to an account of consciousness. Although at the end of **A Kind of Life** Honderich seemed to suggest some rapprochement with the idea of freedom, however, early on in **Actual Consciousness** there are hints he has gotten over this (perhaps the inclination was no more than a Humean sympathy). The argument against **Honderich**, if this is the way he proceeds, must be that the commonplace language of consciousness is not figurative but a perfectly adequate way of talking about what is real. The normal language of consciousness is about the actual nature of what is physical.

The introduction also prefigures how Honderich wants to work through the problems. He will start with common-sense notions of consciousness, notions with which we will all agree. He then proceeds to impose on common-sense deeply embedded theoretical constructs from the history of Western philosophy. Thus what has to be considered, as though this is obviously to be talking about consciousness, are, the consciousness of perception, the consciousness of thinking (*reason*), and the consciousness of wanting and desire (*passions*). No more and no less than **Kant**'s aesthetic, analytic and practical reason, or even standard structures within Empiricism. **But this is terrible**. These analytical devices are so simplistic and in no way address the realities of consciousness. What is needed is fusion, and a fusion of so many more elements than this tripartite division has any grasp of. Already the approach is heading completely in the wrong direction, despite constantly urging readers to attend closely to their own experience of consciousness.

Because *actuality* is not immediately transparent the early questions (in the introduction) must be cryptic. These questions are '*What is it that is actual*' and '*What is it for it **to be actual**'?*

Honderich uses the phrase *affectively conscious* to name what he takes to be a part of consciousness, i.e. wanting, desiring, and pointing to its own specialism, moral philosophy. What is going on here is probably an attempt at a certain comprehensiveness following on from registering what is independent of consciousness (external causality), then thinking about it, and then interacting with it (the **agency** of consciousness in the world, i.e. not just the **being** of consciousness). So how **we** do things in the world, how *passions* (the archaic philosophical model) drive. This though is reductive, simplistic. Analytical depth being derived from an inadequate grid, an imposition, tripartite isolation.

It is not as though **Honderich** lacks a sense of what he would think of as the complex facts of consciousness, but what are in reality the glaringly, self-evident facts of consciousness. He does keep attempting to broaden the parameters. So he allows there may be many more parts to consciousness than the three 'commonsense' parts with which he wishes to open the discussion. And he does keep appealing to the reader not to lose sight of ordinary understanding of consciousness, suggesting that as individual conscious beings we are all experts about this subject matter. And he does point to persuasive, basic dimensions like consciousness being the contrast with not being conscious, e.g. (his examples) dreamless sleep and being knocked out. But the fact is that we already possess a perfectly adequately language of consciousness for understanding what it is. It cannot be reached if everything is proceeding in the wrong direction.

The initial separation of perception, thought and desire as commonsense modes of consciousness fails to see how seamless consciousness is and is experienced as being. For instance, at an alarmingly simple level, and this idea is virtually Kantian, perception without thought is nothing at all.

(Insert. While reading the ***Honderich*** *I became aware of the following: from Ian Tattersall's review of Pat Shipman's* ***Invaders - How humans and their dogs drove Neanderthals to extinction-****, Harvard, TLS May 1st 2015. "...* ***what makes the vanishing of the distinctive Neanderthals particularly poignant is that they were so evidently skilled, resourceful and highly sentient beings, with brains as big as ours. ... indications are that they did not process information exactly as we do and almost certainly lacked language, our single most striking mode of cognitive expression."*** *Therefore clear affirmation of thought without language. You do not need a language to think in. To think about the world we do not need to simulate the world. We think about the world directly,* ***because we think and it is there.*** *Correspondence is a species of thought not its genus.)*

Preamble over, **Honderich** starts addressing theories of philosophers and scientists. Starting with **qualia** (singular). Qualia = *the way things seem to us (****Dennett****), the way coffee smells (****Thomas Nagel****), colour sensations, pains, thoughts on the tip of the tongue, musical experiences (****Dave Chalmers****), 'The problem of consciousness is identical with the problem of qualia, because conscious states are qualitative states right down to the ground. Take away qualia and there is nothing else' (****Searle****).* But shouldn't the initial reaction be that progress is not going to be made if we start by confusing what we are conscious of with our being conscious of it.

Thus the way coffee smells, the pain we experience are what we are conscious of. Perhaps **Searle** gives the game away by being aghast at the notion that consciousness is nothing if it is not those things of which we are conscious. But consciousness, in one sense is nothing at all. This sense being that it is the stream in which fish swim or a screen awaiting images: **flowing** or **on**, unless unconscious.

Honderich sets the topic of qualia within the traditions of Western Philosophy. **Descartes, Locke**, **Berkeley, Hume**, **Kant, Russell**, **Ayer** (idea, impression, sensation, representation, sense-data). And in the discussion **Honderich** connects these concepts with *seemings*. He uses the concept of *truthfully apparent*. All of this really is the discourse of empirical scepticism and verificationism.

...our aim is a theory of consciousness, which will at least have to include an answer to the question of whether being conscious is a physical fact ... This may or may not be promising. If consciousness is to be identified with neural events then not promising, but if consciousness is taken as a physical manifestation in its own right then promising. That there is *promise* is suggested by *I'd say we need an idea of the physical itself a very great deal better than what has been made use of in existing theories of consciousness.*

He then moves on to *subjectivity* exploring whether an account of this reveals what consciousness is. This involves some notion of the self i.e. some unity bundling disparate experience, and some notion of the indubitability of finding oneself a thinking thing. On the whole I think he is supposing that resolving these issues of subjectivity is secondary to knowing what consciousness is, what it is to be conscious. With regard to issues of the self he references **Galen Strawson** and **Aaron Sloman** as standing on different sides of the divide, but there is no detail. The main dismissive criticism is one of circularity. For instance equating consciousness with subjectivity is as revealing as equating subjectivity with consciousness. Neither advances an analysis of consciousness.

Intentional objects. The notion that consciousness **is of** or **concerns something** -its conceptual structure- and that this cannot be equated with some independently existing thing because what one is conscious of may not exist (the example given being **my glass of wine of which I am thinking despite someone having drunk it in the meantime**). This leads to positing some quasi object, the intentional object, as though what one is conscious of has to have some mode of existence. What one is conscious of is an intentional object. But this fundamentally misunderstands thinking. We can think of what is not the case just as we can think of what is the case. In both cases there is not some kind of object of which I'm thinking, I am in the one case thinking of something that is not the case and in the other something that is. There is no mental entity which is my glass of wine which is no more, insisting otherwise is to invent representations, as though consciousness has to have **objects** in some sense stronger than consciousness entailing consciousness of **(grammatical object)**.

The position here criticised (intentionality) is summarised by Honderich thus **States or episodes of consciousness have internal objects or contents that do not exist in the ordinary way of physical things but rather in their own intentional or mental way** and then **All consciousness consists in representations of things**.

I suppose from here the temptation is to suppose that a theory of intentional objects is a theory of consciousness because all that consciousness consists of is a world of intentional objects, as though there is nothing to it but intentional objects, **but this fails to understand that consciousness *is* consciousness *of* and that what it is of is not consciousness. Consciousness, on the other hand, is contentless, in a sense nothing at all, just awareness.** And what is that? I think Honderich moves in the same direction; towards the same question.

He then deals with phenomenality (Chalmers). This adds little.

Section **3** is **Something's Being Actual**. Worryingly the enterprise now declares itself to be essentialist. The language of necessary and sufficient conditions is invoked. The 'five leading ideas' (qualia, something it is like to be that thing, subjectivity, intentionality, phenomenality) may exhibit some common characteristic and it may be this that will allow purchase on ordinary consciousness. A contender is that being conscious is *having something*. Thus,

It is that being conscious is something's ***somehow being had****. So it is* ***something's somehow being held, possessed or owned****. Consciousness has that characteristic.*

I find the apostrophe here puzzling. If consciousness requires that something be had or held or possessed or owned, then whatever the something is is not consciousness because consciousness is consciousness of that something. Or is Honderich after consciousness requiring something having or holding or possessing or owning, or is consciousness having or holding or possessing or owning. Perhaps he only means that consciousness requires a concept of **mine**, or, perhaps the same thing, consciousness is necessarily possessive. Another dimension here will be the possibility of collective consciousness.

I think what I want to say is that consciousness is something's, an attribute of something but that in itself it lacks content, and, as consciousness, from one conscious thing to another its only identity is numerical, and all that that implies. Hazardous I know, but like the cinema screen, identical from one cinema to another, or can be, but populated by a variety. I'm playing around the margins of monads here, and so this a reminder of **Leibniz.**

But **Honderich**'s sense of **something being had** becomes much more dangerous as he progresses. He wants to distinguish it from 'your having consciousness' which he is saying is just being conscious, instead he wants to direct attention to what I will call **something being had in consciousness** i.e. not the something of which you are conscious, e.g. the cloud which looks like a camel (in this case something independent of consciousness). Honderich hovers towards phenomena, qualia, sense data, intentional objects, i.e. something between being conscious and, in the example I have chosen, something independent of that consciousness, something mediating and which is had. Of course if another example is chosen, for instance, having a mental image then something being had prima facie seems more plausible. I will argue this is not so in so far as mental images can be said to be had but saying what this involves does not help with the concept of consciousness. Perhaps we can say some things only can exist in consciousness, but they are distinguishable from the consciousness in which they exist. The cinema image exists on the cinema screen but is not the screen on which it exists. However, the analogy is much less than perfect if only because, e.g. in the case of the camel in the cloud, being conscious of this must involve a thought of this, something entirely absent from the screen and the image it carries. We might say the thought of this is the being conscious of it and the thought is something we have, own, possess. So being conscious is having something. But why is this a mediation? Isn't the thought of seeing it as a camel the same thing as being conscious of seeing it as a camel. The consciousness of x is something we have, something that belongs, something which is mine, his, hers, ours. Describing *the way coffee smells* is not a description of consciousness, it is a property of coffee. Describing the *heifer lowing at the skies* is not a description of consciousness, it is an image in **Keats** poem, and maybe **what** I picture in my head. The treatment I am after is one of avoiding introducing representation between subject and object, as though this solves anything. If one can be conscious of a representation then one can be conscious of anything and so representation is superfluous! And, of course, what one has to be wary of here is the way epistemology is a red-herring.

Honderich distinguishes between having consciousness, a property you have and what you have when you have consciousness e.g. an experience, a feeling, a thought. This he thinks is having something in a different sense. But why isn't each though just an instance of having consciousness? I feel the danger here when H goes on to say that **Certainly something's being had will be related to external causes**, as though there are to be three things, being conscious, something one has in consciousness, and the cause of that occurrence. And this of course concurs with the history in philosophy, a misconceived subject.

H continues to add to his minimalist abstraction, as though this clarifies his subject, so to consciousness as **something's being had** he adds, as **experience,** as **some kind of awareness** and then as **something for or to something else** (this latter returning to **Nagel** and what is x like for x or to x) and then involving having a privileged epistemic position (knowing better than anyone else i.e. in the standard case). Epistemic access seems to be just self reflexiveness (and this chimes with the essential privacy of consciousness, which is an idea to develop). The fact of experience generates another fact, knowledge of that experience. (And remember this is early on in H's account only 14% read to date.) H says at this stage there is no theory of consciousness only a gathering of essential characteristics. It looks too as if he will want to deconstruct epistemic access through these notions of **for** and **to something**, including e.g **the room as it is for someone**. (But I will want to say that in the straightforward case the room **is** all the different ways it is for someone(s).) H is setting up (I'm not saying he is aware of this) a divorce between reality, the world experienced and the experiencing of it.

Then we get something **great** with H referencing **G.E Moore**! The problem for Moore is **the blue of the flower**. H's paraphrase, *If we try to fix our attention on the consciousness rather than the blue ... the consciousness seems to be **transparent** or **diaphanous,*** *to* ***vanish, to be mere emptiness - it seems all we can find to fix our attention on is the blue.*** For me this is the starting point for consciousness, whereas for others consciousness is therefore nothing, and instead all there is is the world.

H then constructs a list, what he wants to call a database, which is a list of the things he has emphasised, like *consciousness is something's being had* (the list **Kindle** *location 1972-2014).* He then says that ordinary consciousness having these listed characteristics is ***actual consciousness, consciousness as something's being actual, consciousness as the actuality of something***.

In a way this sounds empty. E.g. what is an actual banana over and above a banana? Perhaps to be contrasted with a plastic banana. But there may be something here. As I recall H wishes to distinguish between this property and properties like *being tall* (I think he uses the word **protean** in this context). Would it be analogous to compare the properties of a tv, like screen size, being HD ready, led screen etc with the telly **being on**. When something is conscious it is *switched on, the property is actual, on-going, alive.* All of this is metaphor, but there is a promise by H that he will have to deal with *metaphor.* After all, although I want an account which already sits with us and is not deconstructed by scientific narrative, it is not easy to bring the subject into the light, to say what it is we all know. Alongside *actuality* H includes in his list other notions of which he approves, e.g. *something being right there, coming to us straight off*, although *actuality* is the **summary** of which he approves.

Considering his database H introduces **Wittgenstein** and the idea of family resemblance as though this is just something else to think about, but this methodology should be integral to doing philosophy not just a possible add-on or something to bear in mind. H hints that, as he has been informed by Aaron Sloman, the general problem has been faced in particularity in computer science by lines of thought in what is called **polymorphism**. H says his own impulse is to leave it as an open

question whether the *actuality of something* = an effective generalisation or serves as a name or label.
Then something useful (location 2329 **Kindle** edition) (following using **Descartes** and **Locke** to bolster the plausibility about actuality) the etymology, derivation, of the word 'consciousness', namely from *conscire* which is (?) *to be privy to.*

When it comes to considering the intrusion of metaphorical and figurative language in exposition of consciousness H does not retrace to the figurative items on his database and attempt analysis. Instead he just addresses the general question of using metaphor in philosophical and scientific theory, suggesting a methodological preference for metaphorical starting as a sort of setting up of hypotheses. He does not distinguish properly between metaphor and simile but simply talks about how saying what something is like is a way of addressing what it is, and seems to suggest that this is taking something that is clear and holding it analogous to something unclear. This is a matter of using x to reveal y where x is said to be like y, but most often the point of metaphor and simile is just seeing the **likeness** between x and y, it is this which is the discovery, the thing not seen before. It is not clear from his discussion that anything on his list is metaphorical or for that matter not metaphorical. And this is a problem with the whole. Cases are not examined in any concrete detail.

So much of the H is a constant preamble and in keeping with this is a brief synopsis of existing theory plus the promise of what's to come, **'the theory to come'**, of which at this stage (19%) only reveals itself as consciousness and actuality. Very roughly he is shaping up to consider various formulations of the **mind/body problem**.

Then something to make you wonder about the competence for the whole enterprise. In considering dualism H thinks it flawed because it has no explanation as to how something non-physical can cause or be caused by things physical (e.g. thoughts and feeling cause lips and limbs to move and thoughts and feeling are the effects of physical things). But on what grounds does a dualist have to think these connections are causal? This seems a total lack of sophistication in exploring the complexity of relations. My remarks here are not a defence of dualism, only an objection to supposing the relation of e.g. thought to action or object of perception to perception is causal. Once you think the relation **has** to be causal then the dualist thesis is in trouble and in the kind of trouble H thinks it in. However, it is odd to say the apple laden tree causes me to see it, or thinking about lifting my little finger causes it to rise. However, another case: **touching** may be a fundamental case for consciousness: **William Harvey** and his investigations of the '*capering bloody point*' at the centre of fertilised eggs, whereby probing produced '*shrinking away*' '*as though they took it unkindly*'. (**Wilhelm Reich will add depth here reawakening the mechanist/vitalist debate about the origins of life. Is there a life force that distinguishes organic from nonorganic matter or does all matter follow the same physical and chemical laws. All of this leads to Reich's research programme on the electric properties of the skin and the erogenous zones and the attempts to free libidinal energies fixed in the body's 'muscular armoring'. Beyond this Reich 'discovers'** ***bions,*** **bacteria-like bodies which are the missing link between living and nonliving matter, pulsating transitional particles that bear on the origins of life.**) Certainly we would say that being touched causes us to notice, or being pricked with a needle causes us pain. But to say, as H does, that thoughts and feeling cause lip and limb movements is insensitive to how a smile or kicking **are** our consciousness. At the same time there seems nothing wrong with saying consciousness has '*causal efficacy*'. This is just the point that consciousness changes the world, and changes the world because it is part of the physical order (this the denial of dualism). Flesh and blood are living matter.

H goes on though to make pertinent points. Any particular consciousness has a history, a start and end point in space and time.

It is therefore a reality *'a fact as large as any other'.* This then is what I want: an obvious, straightforward fact but mystified by science and theory.

Thought about Zombie (Kirk). It is just the problem of the *imitation game.* Imitates everything about a conscious being but does not have consciousness. So transposed the problem in IT is *what do we need to say if a machine is conscious, or what more do we need if it is able to do everything we do?*

H introduces theories of **multiple realizability** and **type physicalism** (part of his discussion of *functionalism*). **m.r.** supposes a type of *mental state* may correlate with a range of physical states, thus pain may be experienced by humans and monkeys but the physical states to be correlated with the instances of pain may vary. **t.p.** on the other hand is supposing that invariably correlated are a physical state and mental state and that they are both correlated properties of the same thing (this gives rise to the appropriateness of the word **monism**, as distinct from **dualism**). It seems to me an empirical question as to whether a particular state of consciousness coincides with other different or other identical physical states, but I do want consciousness as a property of matter rather than consciousness and matter being properties of something else, (what could this be?!). I suspect epiphenomenalism starts to creep in here. But then is it that simple, because if you say consciousness is a property of a physical thing, i.e. one other of its properties, is it not that the physical thing is a correlation of its properties, which I suppose is **t.p.?**

In H's treatment of *abstract functionalism* we get **'We all knew that a feeling of pleasure is an effect and a cause'**. This is conceding to abstract functionalism but going on to make the point that this fails to tell us about the nature of consciousness. I point to my previous remarks, and H has already conceded that causal relations are law-like i.e. whenever C then E. All of this seems loose.

Good phrase '*covert metaphysics in science*'. c2925 Kindle.

Then H in being what he thinks is thorough is giving **functionalism** another go (because of its links with science and because of according to Honderich the overwhelming success of science) and pursues **Block** some more. But we find **Block** saying such things as '*What is there common to all pains in virtue of which they are pains?*' and I despair!

H has gone through this, **his example of consciousness**, in real time, whereby there is a wood, then glimpsed through a window, then filled with pleasure at the sight, then forming the intention to commune more directly, then opening the backdoor to leave the house for the wood. And he relates this to C, types of conscious states, c tokens C, N, types of neural states and n tokens of N. Suggesting that this is the real stuff of consciousness rather than thinking of pain in general. This the paradigm, what we need to get hold of, taking it that he has been sensitive to the reality of consciousness in his description. But his account is individuation and a travesty. So he wants the particular *feeling of pleasure* as a discreet, individuated c and I think part of a causal story (his predilection) but there is no notion here of how the experience is a coalescing and so unavailable to analytical treatment. The perception of the wood is a pleasurable perception filled with a longing to commune spilling into making for the backdoor without any individuated intention to leave the house as a separate causal event, yet the movement to the door is fully intentional. I think it clear H supposes that there is some discreet, individuated moment of consciousness which is **his feeling of pleasure** at the sight of the early morning wood. As though there are these distinct temporal events, c1 sight/perception of the wood, feeling c2 pleasure at the sight. But there is no feeling of pleasure in itself, the pleasure is in the perception and bound to the sensed propensity to be in the landscape perceived. What is needed for consciousness is a realisation of fusion, what is needed is **life**, not a dead causal chain of physical events. H **though** does wander back on point in reminding that the dawn of which he speaks is always unique and so, I suppose, cannot be linked causally with c. There can be nothing law-like here.

Functionalism is denying content to consciousness, what we have instead is something which is no more than an effect and a cause. This is not very clear until we start to use words like tendency and disposition. Therefore **pain** is an effect and what is this effect, it is a cause of avoidance behaviour. That's all, it has no further content. Then H concedes that functionalism can be an adequate account of unconscious mentality. The problem for it is that it does not have grounds for a distinction between conscious and unconscious mentality. Also, what has to be considered is that unconscious mentality is an empiricist metaphysics, in so far as the idea of sudden, spontaneous, conscious insight, what consciousness is capable of, cannot be allowed because outcomes have to be the result of pedestrian, causal process, a sort of law of physics. What is suppressed here is the magical property of physical consciousness.

But, perhaps there are difficulties here for what I want to say; brought forward in considering functionalism. I am resisting causal accounts of the relations between connected conscious states or of different moments in a conscious process, claiming such explanation is crude, insensitive. However, if consciousness is a property of what's physical (and I'm hesitant this is the formulation to use, but suits for now) will it not be that that which is physical will cohere with the causal order of things and then where does this leave what is conscious, or the determinations of conscious process. It cannot be that a physical identity is just a thing. It has to be a physical process and will not the physical events of this process be causally connected? Am I setting up for myself the task of rewriting how the physical world is in order to get rid of causality from conscious process? The answer to this must be that I am. Physicality cannot be left to the physicists.

Beyond functionalism H goes on to **non-physical or dualistic supervenience**. This expressed as:

A conscious state or event, which is not physical, or anyway may not be such, supervenes on or depends on a neural or otherwise physical base within the person or whatever is conscious.

My problems here are **1** *not physical* and **2** *depends on*, perhaps *supervenes on*. My point is that consciousness is physical but that does not imply any reduction in what is ordinarily understood to be implied by consciousness, it is just broadening understanding of what is physical, where physical is otherwise incarcerated in science. Perhaps I am saying physics is a narrow understanding of the physical. A broadened understanding would not want a word like *depends*. Properties of p are what p is.

This is worrying in H ... `*with respect to consciousness and the causation by whatever route to what can be called its external effects -lip and leg movements for a start.*' Perhaps H's whole enterprise is doomed. Lip and leg movements **are conscious**.

And then more that is worrying: '*How can what has no physical properties be the lawful correlate of what has only such properties?*' The presupposition of *lawful correlate* being something all can agree to is the problem. It is an insistence on the centrality of causality! One suspects this will be bedrock throughout.

To add to the anxieties: '*It's exactly my visual experience of the light as red that ordinarily* ***makes me*** *stop the car isn't it? It's exactly my wanting a glass of wine while watching the news at 7 that* ***makes*** *those things happen.*'

Then we have the absurdity of **Davidson**'s **Anomalous Monism** and H is askew in his approach to it. I suspect H wants determinism and consciousness as physical and bound into lawful connection. I want consciousness as physical but removed from determinist, lawful connection. Davidson wants mental events to be mental and physical and the causal order to hold between the physical events. Thus, the desire to sink the Bismark goes with some neural event and the physical sinking of the Bismarck is caused by, in some lawful connection with the neural event. It is through this connection that it can be said that the desire was causal with respect to the actual sinking. **But the mental is physical**. It is the narrowness of science's understanding of the physical that will obscure how this is meant. To say the mental is physical purports to enlarge understanding of the physical.

For H **causal** has to imply **lawful connection**. I think he wants consciousness to be a **reality** and means this in the sense that it has to be causal and this has to imply lawful connection, although he is allowing equivocation over what lawful connection is.

In considering **mentalism** H presents dispositional belief in the same way as anything having a disposition, and his understanding of this is for example, and his example, that fragility is a disposition possessed by cups which with the addition of another event cups break. So that a dispositional belief (e.g. that clocks tell the time), which exists as a **neural fact of complexity,** with the addition of something else precipitates an effect, e.g. an intentional action (looking at the clock). H, then wants to use the conceptual structure of sufficient condition. At this point there is no distinction between empirical cause and logical necessity or any of the other possible variants.

H moves to **Naturalism**. *'one will not countenance any form of explanation that will not stand the scrutiny of scientific and other well-established, pragmatically fruitful methods of communal checking and testing.'* ***Burge*** *'ultimately nothing resists explanation by the natural sciences'* ***Blackburn***. Another implication here is that whatever is explained will fall under the concept of lawful connection. In no sense do I want to diminish the material, physical world, i.e. the world science is concerned with. I do not want to claim there is any other world. I don't want any non-physical world. But what I want is a recognition that the physical sciences only reveal part of this world, that quantification (the measurement) of the physical is like a digital representation, a simulation, not the totality. The ordinary language of consciousness is itself a coherent totality and is empirical, subject both to communal checking and testing, something constantly employed in our dealing with interpersonal reality. Consciousness is an irreducible part of the physical. It has its own language, which is fully coherent, we have no difficulty with it, something empirically transparent and not something mysterious. Although this is not to say our understanding of consciousness does not contain problems and mysteries. But generally speaking it is something self-evident.

Then we get *'a leading idea in our philosophy and consciousness', 'a theory'* is
'All conscious states and properties or properties consists in representations, maybe items from a language of thought or more languages, making up propositional attitudes, these understood naturalistically.' L3736

Added to this we get 'aboutness'.

My thought is that we can think **about** x without there being any representation of x at all and that all the mistakes start here from not seeing how this is a possibility and is not an impossibility. Whereas H is saying that what is indubitable is that somehow representation is a fact of consciousness, not of all consciousness but something that reveals a chunk of ordinary consciousness. **Instead we might say that representation is the means of communicating what we are conscious of.**

H very briefly deals with **panpsychism, double aspect theory** and **neutral monism**. It is his view that they have (these theories) '*so little in them*'. And that they do not pursue what they are after using *clarity, consistency, validity, completeness and* *generality*, which for H is what philosophy aspires to. He then brushes everything under the carpet with the question '*Should I do more reading?*' but implying it would be a deviation to do so. However given the length at which **Galen Strawson** ('*a strong philosopher*') has written about panpsychism it suggests a prejudice against on H's part. After all he is sort of writing a text book on consciousness rather than just an original work of philosophy and his enterprise does not seem curtailed by considerations of length. But the few words of summary H uses suggest interest. Thus **panpsychism** is held to say that *a conscious state is a special case of an inner aspect of every physical thing ... a universal sentience.* And **double aspect** saying that consciousness is one of two **aspects** of a thing the other being physical. **Neutral monism** is said to hold that '*A conscious state is a different arrangement or organization of the same neutral or primitive elements that also compose a physical state.*' These formulations seem like moves to grapple with what has to be faced. All I have at present is that consciousness is a property of physical things. I need a lot more.

However, H moves on to consider **physicalism** (known in the past as **materialism**) which according to H holds that conscious states are identical with states of our brain, so that conscious states are physical states. He then makes the good point that when you press these theories for explication of the physical, i.e. what is meant by physical, there is very little to find: at best the physical seems to reduce to something not very helpful namely the physical is the *scientific*. This then is the problem because what is being said is that the physical is nothing more than what can be conceptualised in science. My pressing question is 'Can't the physical be much more than this?' We understand the physical perfectly well everyday in interacting with the world and very little of this requires the employment of any scientific concepts, although the concepts of science may have their roots in ordinary language.

To H's question to Susan Greenfield and others **L3855** 'Is your consciousness inside your head?' I am inclined to assert differently that consciousness is a property of the body, e.g. one's foot is conscious, it is not as though the consciousness of the foot is in the head nor that the consciousness of the foot is to be located in the head. To think any of these latter things is to have the wrong model of conscious life.

Searle and **biological naturalism,** namely that consciousness is not mysterious, it is just higher level states of the brain realised in and caused by lower level neurobiological elements, just as the liquidity of water is a higher level state of an organisation of molecules. This is property dualism in the sense (defensible according to H) that it asserts consciousness to be a natural reality different from the rest of what is natural. H will run with this in so far as it asserts that consciousness is *a natural thing but is different from other natural things.* This though fails to tell us what consciousness is, apart from it being in relation to the rest of the brain, namely the 'lower' elements of the brain. I would add to this thought (which is H's) that the concentration on the brain is misleading, as though all consciousness is is a higher state of the brain. *Higher & lower* here are worrying too. As though the distinction tells us something, when in fact it just suggests homo-centric values. What H accepts however as part of his fulsome tribute to Searle is that **consciousness is natural in its own way**.

H then meanders to **Union Theory**, which he seems to want, *namely that a conscious event and a simultaneous neural event correlated with it are a single effect such that neither can be caused without the other and they form a sequence with antecedent (previous) pairs.* A problem here is *simultaneous* and *correlated.* What does correlated add to simultaneous? How specific is the identity of a neural event paired with a conscious event. And there seems something wrong with approaching consciousness as a sequence of conscious events. Crudely the counter concept might be **stream of consciousness**.

Section 6 (L4180) is ***What is it to be objectively physical?*** This is something I have to deal with. Making consciousness at one with the physical but not so as to reduce it to nothingness. Is there already a problem in that H qualifies physical with what is *objectively* so? Is this like adding **actual** to consciousness? In this section he addresses, so he says, not physicality in general but physicality as treated by science and philosophy. This then is already to have truncated the subject. If consciousness is physical what is not physical? Do I want to say **that which is not so**? **Remember in reference to a person we distinguish between physical and mental ailments.** H then makes the good point that if physicality and consciousness take their meanings from mutual exclusiveness then neither is clear without a good understanding of each other and that from the position of most who contrast the two but have little time for the latter there is no understanding of the latter and so no understanding of the former. Also he points out that 'decent' dictionaries do contrast the physical with the mental and as they reflect facts about common language this should give us doubts about whether consciousness can be physical, if the distinction *is written into our common language.*

The Austin quote (L4346) is worth holding onto.
... `our common stock of words embodies all the distinctions men have found worth drawing, and the connection they have found worth marking, in the lifetime of many generations: these surely are likely to be more numerous, more sound, since they have stood up to the long test of survival of the fittest, and more subtle, at least in all ordinary and reasonable practical matters, than any that you or I are likely to think up in our armchair of an afternoon - the most famous alternative method.'

This is taken from **A Plea For Excuses**. In a general form this is the point to be made about 'consciousness' in ordinary language. In ordinary discourse 'consciousness' is not a hard problem.

In my materialism, and I do want to think of myself as a materialist, do I want to deny that all physical properties are related to each other deterministically? I suppose I am saying that consciousness is the exception to the rule, which is rather lame. Am I saying what is material is both bound and unbound? H is much more exacting in so far as our descriptions of the ordinary physical world go far beyond the world described by physics. Early physics recognised only material particles, the world of atoms. If this was all we could describe the world we think we inhabit would be unintelligible to us, and, so for most of us, would be the world physics describes. Contemporary physics has anyway replaced the particles of C17th & C18th materialisms. So what there *really is* is even more remote. And so I need to remark for myself that in **IC&VM** the concept of materialism is not really examined, not exactingly anyway, and for **Consciousness** it needs addressing. To this extent H is on the right trail. In schematic terms, and so misleading terms, I suppose I want to say that if where C1 and P1 are conjoined and then there is C2 and P2 and if these are genuinely conjoined then P1 is not the cause, does not determine in a causal sense P2 (C for consciousness, P for physicality). So that the conjoining takes what we are dealing with out of law-like relations. Instead we have **autonomy**.

A hunch is that as what H is calling the microcosm that contemporary physics addresses leaves a degree of uncertainty or indeterminacy about a law-bound causal order he may want to sideline those theories in favour of a discourse, preferred as central and ordinary, of dimensions we all inhabit and recognise; a macro-physical, causally ordered reality. If this is the inclination then things certainly are going wrong and suggest an inability to get to grips with quantum theory.

(Something that strikes me here as the discussions get closer to physics is that both General Relativity and Quantum Theory do not admit consciousness into their defining equations, and if consciousness is part of the nature of the physical, i.e. not something outside it and so other-worldly, then any equations without consciousness as a term cannot define the physical and will produce a distorted account of reality. This has to be reconciled with the undoubted internal consistencies and inconsistencies in the relevant equations and the undoubted empirical corroborations and falsifications of these equations. In addition there is the question of consciousness and the limits of mathematics. If consciousness as physical can contravene logic e.g. irrational behaviour, then mathematical physics can never do better than approximate to reality and will always be a falsification of it. The point is that it is at least true of human consciousness that we can always act to the contrary, against our best interests and even when we know this, so what we do is what we do, it could not be otherwise, but we can always do other than what we do, ie. act perversely. In which case what we do is not the effect of a cause in the scientific sense. What is required is a wider understanding of what is physical.)

Then H says something that makes me feel that I am indeed Wittgensteinian, namely that to understand the physical we do not need to look to physics but instead to examples taken from ordinary life, perhaps paradigmatic examples, then we can proceed. I want to talk of the rich understanding of consciousness and similarly with what is physical which is built into our ordinary language (the Austinian point). On what basis should we concede to physics when at the level of ordinary language what it says is paradoxical and apparently confused. Physicists are constantly saying that what they say runs counter to common sense but that reality is not understood by common sense, instead its nature is revealed by ... and then we get a mathematical equation, like a rabbit in the magician's hat and hey presto there is no further argument. The problem as I have argued in **IC&VM** is that the mathematics may be sound, although given the contentious nature of the profession this is questionable, but any understanding of reality requires the mathematics to be translated into ordinary language propositions, and at this level of meaning they fail to convince and instead run counter to sense and can only be defended apparently by the mathematical equation of which they are supposed to be the translation. So we might say that if empirically atomic clocks tell different times at different altitudes this might only show that they are affected differently by different altitudes and not that theoretically time itself is relative. The so called proof of the relativity of time is in mathematics, but that this is a proof of the relativity of time is a translation into ordinary language from the calculations, and is not unambiguously corroborated because it predicts the results of empirical experiments which could be explained without conceding Newtonian time. This example I use speculatively and probably requires revisiting. Reflecting on the Wittgensteinian approach H becomes essentialist in voicing his opposition, so examples are instructive but what is needed is a general concept of 'table' ('something to put things on' ??), the physical, consciousness etc, 'surely the very stuff of inquiry'. And again we get a lack of imagination as H suggests that lists of examples for things falling under terms have to have more to them than say listings in a telephone directory, but, the more for H, I am sure, is some logical definitional binding, and so he is far removed from the way classification is part of social purpose and practice and distorted by this (you might say the ideology of classification is that members of classes are bound together rationally through similarity to each other, so hiding membership as a social act of putting things together for social control).

Stuttering on with physicality H suggests we need to attend to **scientific method, to space and time and to lawful dependencies,** none of which in themselves close the relevant issues but are serious attempts to grapple with what is to be unravelled. At the same time H protests that his book is not the place to elaborate these discussions. The subjects are at least as vast as consciousness itself, but despite the omissions progress is still possible, though it needs to be mindful of these adjacent dimensions and how they have a prime place in bearing on the problems. However **lawful dependence** and H's commitment to it is, for me, a parting of the ways. So he says *Anything not in causal or in other lawful connections is merely the mystery of being an abstract object or other dubious sort of thing.* Perhaps as a digression, he does not think the issues it raises bear on the progress he wishes to make, H dips into Hume and regularity or constant conjunction as exposition of causality, but points out its insufficiency when confronted with examples like the regularity of night following day. In its place H puts *the connection between a causal circumstance and its effect* to be *that **if or since the first happens, whatever else is the case, the other happens***. But this won't do. Already he has insisted on the distinction between logical and causal necessity, but the only thing that makes his formulation true is the logical necessity that if anything is to count as a cause it must have an effect. The only way that it can be that a set of circumstances have a certain effect whatever else is the case is if those circumstances are causal, and then, of course they must have an effect because the concept of a cause is such that nothing can be a cause unless it has an effect.

H then draws attention to **lawful, i.e. regular, non-causal connections.** These being further aspects of the physical. The examples given are invariable proportions of volume and temperature of gases at constant pressure or the way they vary together. On a par with this as a scientific example H mentions the *simultaneous rise and fall of the opposite ends of a see-saw*. Volume and temperature of gases are not causally connected, nor are the opposite ends of the see-saw. Put simply this seems to be a reference to the conjunction of properties and I suppose one usual distinguishing feature between property conjunction and causal regularity is simultaneity in the one case and antecedent/postcedent conjunction in the other. So we might say simultaneity rules out the position of one end of the see-saw being the cause of the position of the other end. Although there is something suspect about *lawful connection* between the ends of the see-saw.

(H mentions **Ayer** and *direct realism* or *naive realism*, i.e. that perception is a direct relation to ordinary physical things. And this reference just after a return to qualia.)

(H gives an airing to primary and secondary qualities in connection with physicality. Points out that mass as much as colour has different appearances from different points of view. This is not intended by him to mean that p and s qualities are only ideas. He is after something else.)

(L4990: *'Let us refer to this composite fact as the fact that objectively physical things are separate from consciousness. They have the characteristic of* ***separateness*** *from consciousness.'* And objectively physical thing is being considered under occupying space and time and being in lawful connection. And that what is objectively physical is independent of knowledge of it.)

Something that next emerges is that physical properties are not such that only *one particular individual can be conscious with respect to it.* This is the point that what is objectively physical necessarily eludes solipsism, it is this which makes it objective. And this seems a good verificationist point, but if we are to argue that consciousness is a physical property then on what grounds can a moment of consciousness be objective. Here we are exploring the necessary secretiveness of the mind and how can this be so if consciousness is a physical property, surely it must be objectively accessible. H mirrors these worries, saying *each of us having a greater confidence about our own consciousness than others*. Firmed up in L5015 ***'Physical properties are objective in not being a matter of someone's or something's privileged access.'*** And that physical properties are uniquely open to perception, perhaps open to more than one sense (**Ayer**). What I have to argue here is for a broadening of the physical. So H summarises Searle as saying that physical properties are those whose existence and nature is reported in truth and logic and not owed to feeling desire e.g. having to do with personal relations, **but** the Srebrenica massacre is a mass of physical properties which cannot be understood as the fact that it is, the physical state that it is, without facts of consciousness, facts of personal relations. H has already conflated ordinary physicality with quantum physicality, conflating sofas and atoms. This is something like the same point.

(H quotes **Sloman *'the self - a bogus concept'*** But consciousness refers what it is conscious of to itself. And what is that? A particular consciousness in a particular space at a particular time, an instance of a physical property. Is this just a point about a self-conscious consciousness?)

But H starts moving in the right direction, suggesting that he is teasing out the limitations to a scientific concept of physicality. So we have *self, unity, freedom, desert, retribution.* Could these be part of physical reality -broadened-. But we get: *Any self or homunculus or other kind of unity inconsistent with the characteristics of the objectively physical that we have, of course is not part of the objective physical world.* Sort of waiting on which way he's going to jump. Then his moves towards *scientific standards of objective physicality make for hesitation about taking consciousness as objectively physical,* and this much can be granted. So is consciousness not physical at all? *Not physical in any sense?* (There is too much deference in H. Those he considers to be substantial philosophers who have addressed the problems have to be allowed a *mouthful* but he does very little with it: he just pays them, and some of the big issues they give rise to, lip service. Is it just that he is trying to convince that he has done his homework? But his homework favours a particular consensus)

H' lists of physicality and objectivity L 5152-5183.

Then Chomsky: there is no possibility of a concept of the physical, Newton's solid matter has come apart.

Then H specifies a programme, in three parts and to come, named the *actualism theory of consciousness* or just *actualism.* This the sequel to the *primary, ordinary sense of actual consciousness.* But then H starts to explore what consciousness is from the point of view of what we are conscious of and this is put into a category distinguishable from the physical world, it is, instead the **subjective physical world**, made up of myriad parts. I think rather I want to say that consciousness is not unravelled by what it is we are conscious of. What we are conscious of is the physical world or much of it is. What has to be answered is what consciousness of it is? **Consciousness is being in touch**. And H's approach here is a concentration of *perceptual consciousness*, marking it off from what he is calling *cognitive and affective consciousness*, although these accompany perceptual consciousness. This though fails to grasp fusion, everything is integrated in consciousness. H cannot outstrip the philosophical tradition, these demarcations are deeply embedded in the philosophical tradition and the subjective physical world and the physical world distinction is in danger of being a rehashing of the phenomenal/noumenal distinction, only with science playing the role of speculative metaphysics. The relevant conceptual structure has to be remade. And it seems H's subjective physical world is to be saved from scientific derision by anchoring itself in a common sense absolute spatial and temporal world alongside relativity theory!

H then attacks a version of representationalism. What I am conscious of is given in consciousness not inferred, deduced, not an *iconic mental representation* (**Fodor**). So H decries the idea of a screen between us and the world. All of this is good and H compares it to Ayer.

Then H seems to want to say that his awareness of perceptual consciousness of his room is not actual, I suppose, not given, but something separate. But for a self conscious consciousness awareness of awareness is part of awareness and so, I suppose, 'actual'. But perhaps H thinks he can isolate something which is perceptual consciousness of his room, this is just the room made actual, not the room but the room made manifest, a presentation, a reality which is not the atomic, sub-atomic world, not representations in the mind, but the world as actual, not a ghost but really manifest, a sort of unsynthesized manifold. But this is just transferring the problem to the objects of consciousness. I want to say this is unproblematic until philosophy and science start interfering. Consciousness encounters its objects, there is a magnetism between the two, perhaps like sexual attraction.

H presses the attack on images (retinal), representations being the objects of perceptual consciousness and does so with good rhetorical force. L 5502. I cannot recall that he has gone out of his way to explain the philosophical compulsions that lead to qualia, sense-data, but it is this needs dismantling rather than asserting the world actually is there for us, although the latter is correct.

Clearly H's points about perceptual consciousness cover the senses not just vision, but he is inclined to compartmentalise rather than connect organisms holistically. Wollheim's influence seems non-existent here. Although he complains of admonitions from Bernard Williams that as a graduate student at UCL he failed to make sufficient distinctions, while studying for a PhD.

(Thinking is not the paraphernalia used to represent it. And what is it we can remember before those times when we acquired language or our language was minimal?)

Is there a tendency for H to build consciousness like an animated film? Discreet shot added to discreet shot, albeit run together too fast to see the separations.

The world itself is as much duck/rabbit as is duck/rabbit, and of course duck/rabbit is in the world. So when H is being facetious about the philosophical pretensions derived from some notion of **Cezanne**-like perception, i.e. seeing the world as a flat dimension of shapes and colours rather than seeing three-dimensional, identifiable objects, he could allow that both perceptions are perceptions of an actual world. The world is duck/rabbit, objectively.

H's exposition of the actual is another instance of the philosopher and his furniture. Instead, what is needed is bringing to life the way a body is met and meets other bodies. The life of the physical. Science gives us the physical without life. **Memory is a dimension of consciousness. It is not that we have consciousness and then we consult memory. Consciousness ⊃ Once Now Next, C ⊃ ONN** for any consciousness.

Then H turns (at last?) to the *argument from illusion*, something to be refuted if progress is to be made. H objects sensibly to **afi**, but does try for a knockout blow which relies on the empiricist creation theory, against which I have strenuously argued elsewhere. I.e. how can there be representations unless there is something we know which the representation is a representation of. H then tackles a variation on the **afi** namely the *argument from hallucination* (the attack on *naive realism* or *direct realism*). H wishes to distinguish *nr* or *dr* from his *actualism*. *Dr* H takes to be saying that seeing something is a direct relation to an objective physical object, and the problem with this for H is that it does not say 'what the relation ... is'.

Perhaps I should say something about hallucinations! H sort of appeals to a distinction between perceiving something and thinking or feeling you are perceiving something. Well that is correct but enough?

Get the feeling H is attempting to construct consciousness a brick at a time, a sort of digital enterprise, like the previous thought about animation. So the start is with perceptual consciousness, a sort of Kantian transcendental aesthetic, percept, space and time. We start with percepts in which the room, the desk, Ingrid's drawings are actual. This is all we have. So what *it is like being me is not now actual with visual consciousness*. But this is distorting abstraction. It is not being able to let go of the history of these discussions. For H these discussions just have to be rebuilt with some changes to the language but not the structure. The idea that a sense of what it is like being me, memories of my seeing are actual in seeing the room, the room as actual, H is denying. This analytic approach will miss what is holistic about consciousness. H even suggests that a holistic account is a matter of being *self-obsessed*! This analysis is also aimed at removing the *metaphysical subject*. And that bit is ok. But visual experience is possessed, albeit by a contentless subject. And H gestures to something like this but paradoxically, given his objections to the metaphysical self, goes too far in its direction as he talks of experience hanging together with something ongoing. Empirically this is a general truth but theoretically not so much is required to make minimal consciousness comprehensible.

He then denies that the room is *seen through some transparent but still visible medium*, something that can be discerned perhaps with difficulty or effort. But empirically is this correct? What do we say of floaters? They are actual but are not external although are 'seen' as seeming actual but despite this are quickly discounted as being external. And what is external objectively is seen through them or around them. And what of migraine flashes?

However, we get '*Of course there isn't just perceptual consciousness going on with me now. There's thinking and feeling going on as well, including believing and reacting with respect to the seeing...... there are funny questions, like the question of whether there is feeling in the thinking.*' So! but then '***So what? I can tell the consciousness of or in the seeing from the thinking and the feeling. I sure can.***' And, '***There really is* nothing *there in between the***

room and me, whatever I am.' And of course there is something here I want to be a part of. Namely when I open my eyes there is the world, unmediated, an objective world, independent of how I am constituted. This should be empiricism. And then for H perceptual consciousness is somehow prior to cognitive and affective consciousness. So a construct of the mind looks in the offing and looks very much as if it will start with a tabula rasa structure: '*the source material which the mind develops.*' **I think I want to say that understanding and consciousness are coexistent**. A phrase H is starting to make use of with increasing frequency is '*subjective physicality*', it is this he is to give an account of.

Half way through the book and signs of mistakes are becoming clearer. H has this distinction between *subjective physical world* and *objective physical world* so that he wants to say that *actuality* is dependent on both, e.g. the room he perceives. And he quotes **Nagel** in this connection '*how things seem to us depends both on the world and our constitution*'. But all of this creates the traditional philosophical gap (noumenal/phenomenal), or at least the same structure. What is needed in opposition to this is that the objective physical world **is** all the ways it is experienced, that it actually exhibits all these appearances, this is how it appears and how it appears is how it is. Of course there are hallucinations, floaters, migraine flashes, etc., but these can be distinguished and so are not how the objective physical world is. H is just storing up trouble in preserving the structure of sense data/objective world and does not avoid this by insisting on physicality on either side of the distinction. The trouble is that he is not far away from **Nagel**'s language, namely '*how things seem to us*'. We are not cut off in seemings, we encounter the world and this is not a matter of starting with something physical in the senses, something temporally and spatially synthesised, something before everything else or at least distinguishable from everything else.

(Worth thinking through comparing hallucinations with dreams. In dreams we seem to see. An hallucination is to be awake and dreaming? Both are forms of consciousness.)

L6190 '*That ordinary belief is that something like an objectively physical desk, and facts about the eyes and brain, and the seen desk, are tied together.*' Is this though **ordinary belief**? 'something like' an objective x? The ordinary belief is that there is a desk and not that there is something else which is the '*seen desk*'. The ordinary belief is that the desk is seen and that we need our eyes for seeing it. Not sure that the brain is part of ordinary belief in the same way. It is not given in ordinary experience although it is ordinary knowledge.

Extra clarity about **actualism** : '*Our actualism ... is to the effect that there is the physical world and it divides into two domains, the objective physical world and the subjective physical world, the latter being in certain relations to some of the objective physical world.*' Is a clue to this given in what precedes, a few paragraphs earlier, where there is a distinction between atoms and sofas? Certainly H wants the objective physical world to be '*within science*' and then there are desks and retinas/neural circuits. The objective physical world and the subjective physical world are for H just **parts** of something, the physical world (L 6259). I would deduce from this that the physical world has parts two of which are objective p w and subjective p w!

'**subjective physical worlds are not in heads**' (L 6275)

(The mystery of consciousness is not 'how can there be consciousness?' 'what is consciousness?' etc., but the fact that consciousness is a secret life and so necessarily a mystery. If I don't tell you even the brain scanner doesn't know.)

'*Subjective physical worlds, further, unlike the objective physical world, are almost always a matter of consciousness of one particular individual perceiver.*' L 6292

H echoes the comments above about mystery, for H 'privacy'. But privacy is said not to be a deep fact but just a consequence of the neural and other uniqueness of individuals. **But consciousness is not just the uniqueness of point of views which, of course, up to a point, can be shared.** And H is prepared to allow that developments in neuroscience might intrude on this privacy. I suppose my fundamental objection to this is that consciousness is misconstrued if it is supposed that there is some code in the brain that can be deciphered into propositions, the propositions of consciousness. Thinking does not require representations. Representation is just a way of communicating thinking. Clearly this is a challenging hypothesis.

'*the fact of subjectivity that is* individuality'. H is flirting with some notion of binding for discreet subjective content. And we get from H the notion of synonymy between cognitive and affective consciousness via cognitive and affective representations. All of this is sounding very much like **percepts** and **concepts** and this very unnatural division of consciousness into its analytic components, all being held together by the superglue of individuality. This isn't progress. This *individuality* H calls '*a successor to a metaphysical self*'.

Prima facie I can accept L6392 **8. Hence a thing in a subjective world is of course different from different points of view.** But a thing in any world is different from different points of view yet is the same thing. But then from H we have **10. Subjective physical worlds .... cannot conceivably be separate from consciousness.** It seems H wants a physical world as revealed by science ('microcosmic') and a physical world of ordinary experience (a bonding of particles and Ingrid's drawing hanging on the wall), but that the latter just is for consciousness, that's how it exists. And you might say what is the world of ordinary experience apart from its being experienced? Is it there without consciousness? If there is no consciousness is there just bundles of particles, the fundamental stuff of being, brute being, not Ingrid's drawing? But then H wants a subjective physical world to be the perceptual consciousness of only one particular perceiver, so that he can talk of **any** subjective physical worlds, i.e. a plurality, not just the one subjective physical world.

L6412 H is saying that there may be objections to his meandering, although he justifies this on the grounds of the complications of the subject, but he recognises there will be demands for generalisation and simplification. So he offers some. **What is actual with perceptual consciousness is worlds, sometimes rooms and they are physical and subjective,** 'consciousness as existence'. So it would seem the room in which his book is being written without H or I or the police entering to investigate their disappearance, is not actual, does not exist. Is this Berkeley? And 'Who says that when they have their eyes open that what's there isn't reality? It's reality in the primary sense, isn't it?' This then is what **actualists** say. There is 'something really out there', 'a spatial world, causal and so on'.

I mention **Berkeley** and his name comes up a few paragraphs later, mentioning Johnson and the stone to be kicked. And then we are told we might be **hooked** on the idea that being perceptually conscious is **about** or **of** something **out there**, 'not the dependent existence of something out there'. Is this addiction aimed at myself, I mean at my own position? **Naive or direct realism**? So I might be thinking that my being perceptually conscious is something being in relation to something out there and so not what H is claiming for actual consciousness which is being perceptually conscious is something's being out there. Or perhaps H's difficulty is that in saying perceptual consciousness is being in relation to something out there, it sounds as if by implication there is something in here, some internalism, and it is this distinction he resists. I don't want to say perceptual consciousness is anything other than a (and this is loose) property of what's physical, so it is a physical property and in so far as I perceive, the perception is of something. Without there being something I am perceptually conscious of I am not perceptually conscious. There is the physical world and part of it is a consciousness of itself. Is this its actuality? Certainly I want no truck with internalism, and no **species-pride, the elevation of ourselves** although I am arguing for 'free-will' (again loose) despite not thinking *we're above nature.* 'Free will' is part of nature. That's not quite right but in the right direction for what I want to argue. H in this regard continues to insist that he is only dealing with perceptual consciousness and not, thinking and wanting, cognitive and affective consciousness, and that different issues with regard to internalism arise with the addition of these other topics. But again this just looks like the artificiality of **Kant**'s aesthetic, i.e. there is just the room, a percept, something physical, something isolated. I want to say with this stripping there is no such thing.

This must help to clarify perceptual consciousness and subjective physical world in H. L6517 '**Do you at least suspect that it is surprising if not worse to take it that physical states of affairs outside of perceivers are partly dependent for their existence on, have a necessary condition in, facts somewhere else, in or about perceivers.**' This he must be saying about **the room**. I mean H's position is that which his *you* is suspicious about. So he goes on to remind his sceptic that there are ordinary facts of causation where e.g. a heat or light source here has the effect of a lighted or heated object there, so how can there be a location problem about a fact about H being a correlate of a fact out there. !!?? And he is saying that the desk being green depends both on it and a perceiver, otherwise, he says, *the science of vision is up the spout* (which, my position in my own *mind* is strong enough to say it is). So there we have it. The battle is joined. What I have to say about **direct perception** is to dismantle this. H is saying that something in the world has a property because of something about the perceiver of it and the perceiver's location. But the perceiver doesn't make the desk green, the desk is green for those perceivers who can see that it is and if there are no such perceivers it still is green and this is compatible with it being black as well. The desk has these appearances. It would seem that in the past H has flirted with making use of '**reality being as it seems to be**' to make his points but now regrets this. If I am saying t**he desk has these appearances** is that close to reality being as it seems? The nuances here require careful treatment. H further opens onto what he is trying for '*Are there really two worlds, and two desks in one of mine right now*' (L6548) presumably meaning an objective physical desk and a subjective physical desk. '*Well, I do and I don't say there are two worlds and desks*' but the notion he constantly returns to which he thinks saves everything is that the physical properties in both, if there are both, are in **lawful connection**. Causality becomes the touchstone of objective worlds. Atoms and greenness this is what H is going on about: the nub of the problem. Everything now is unravelling. So '*the dependence on me neurally of subjective physical properties out there does entail that a subjectively physical room won't be there when I'm going down the stairs*' so I suppose he means what's left of the room is a whirring of atoms when he leaves his room! Something now has gone wrong, this is no longer at one with ordinary understanding as he keeps maintaining his argument is. So we get from H **'Does that make actualism incredible?'** Perhaps this is the best clue to date as to how **actuality** is being used. We are supposed to make the world actual by looking at it, in this sense the objective physical world is not actual, rather, I supposed, it is inferred from traces. H's defence then to it being incredible is that what he is saying is rooted in all human thinking and research into perception, so that all the accounts of seeing make it into something that includes contributions of the perceiver to the perceiving (the science of perception and what philosophy has made of this for that matter). We have here then a shared worry and it is only not a worry for those maintaining **naive realism or direct realism.** But such a position for H is neither clear nor explicit about perception, it just makes an unexplanatory statement that perception is our being directly **in touch** with objective physical things. This though is a metaphor. When we see x we are not literally **in touch** with it. In this context H's knock-out blow is to say naive realism does not have a way of distinguishing perceived effects from unperceived effects, e.g. a circle of light on one's back from someone's little flashlight. But perceiving is like shooting something, and not waiting upon an effect, and you can't shoot the pigeon if your gun is not pointing at it. H pursues a different analogy namely that of the tennis ball coming to rest where the tennis racket isn't, as though we cause the world to be the way it is, instead of our finding, locating the world where it is, as it is, like the bullet hits the pigeon. All of this is very loose, but I'm trying to indicate a difference, which is to be firmed up elsewhere. In contrast H wants to say that *being perceptually conscious is a world existing.*

H then considers abandoning his position for something simpler, namely the unrectangular subjective shape of the desktop, dependent on his point of view whenever, being in fact an instance of the properties of the objective physical desk. This he ties to **J.S. Mill** and a material thing being the 'permanent possibility of sensation' and the empiricist tradition in general from **Locke** to **Ayer**. But this leads to a physical thing just being the possibility of sensation and anyway these sensations, sense-data, qualia etc., don't exist. But **Mill** doesn't mean that a material thing is just the permanent possibility of sensation, but that a material thing with all of its properties including its rectangularity has to be there as a permanent possibility of sensation, or better, and this may not be Mill, has to be there as a permanent possibility of being seen, the realisation of which depending on the existence of perceivers. This excursion by H then runs into a muddle about objective/subjective, and I will not follow him there.

In considering supervenience H sort of stumbles on a stark question, namely, *how can it be that from the soggy grey matter of the brain there arises that technicolour phenomenology of percep-*

tion? What I have to say about this questions or the terms of it, is that a focus on *soggy grey matter* is already theory laden and that the starting point should be **the organism as a whole**. Next phenomenology is theory laden, instead we should be talking of **seeing**. So the question is better put as *how is it that such an organism sees?* Is it helpful here to speak of some brute fact of lawful connection, or should we be asking how it is that physical particulars have properties, one of which is being able to see? And H does go on to raise what seems to me a relevant question and that is, is there a presumption that some likeness exists between cause and effect, in this case between state of the organism and seeing? Does this presumption straightjacket science (H does not raise the question in this way)? But he does point out that many causes are unlike their effects, e.g. chemical process and explosion. Does all of this return us to brute facts, somehow without intelligibility?

'Nagel **not only remarked on the dependence of how things seem to be on both the world and our constitution. He also remarked, as you may remember, that there is more to reality than can be accommodated in the physical conception of reality. He meant more than in the objectively physical conception of it. He was right about that. We have been true to that, but taken it that the additional reality is a second physicality.' 'I add ... that speaking of something as a room is not to diminish its reality but only to distinguish it from other things.'** I think I want to say that a thing, a physical thing is a complex of properties one of which might be being the room.

H then moves on to Cognitive and Affective Consciousness and this is the problem. What is required is an integrated theory, otherwise all that H has said about perceptual consciousness will just sound like another bash at **Kant**'s Aesthetic, i.e. putting the mind together in bits. The point is to get out of this structure, out of a philosophical tradition, more is needed than translation.

Universal Representationism is the name given by H to qualia/ sense data philosophers and scientists who are said to hold that all consciousness is representational to some extent. Also intentionality/**Brentano** etc. But with regard to perceptual consciousness and what is **actual** with it, H is affirming '*What is present in my case right now is a desk, not some sign or symbol of the desk.*'. And this point he is reaffirming 09.05.16 in his **tweets**. With this we concur, although what is present is always much more than this i.e. what I am seeing at any point in time. H raises as questions that universal representationists may have wanted to explain perceptual conscious through representation because they were so sure of representation with regard to cognitive and affective consciousness and that all three are almost always simultaneous. But the tripartite distinction is so disastrous (this my point). Seeing is thinking and feeling. H goes on to talk of the separate problem of attention in perceptual consciousness and this may be ok. At any moment what we see is a manifold and we do pick and choose from this manifold, i.e. attend. **But** H is saying *L6996-7004* that he will be allowing representation *two large places* in cognitive and affective consciousness, in **thinking** and **wanting**. But then a new phrase enters *the actuality of representations*.

Then H becomes promising. We have the written sentence in H's book, well for me the sentence read on my Kindle, which is *The thing's on the way to an end* (he is referring to his book), but which started as a thought on H's awakening on the morning he added this sentence to his book, or so he says. H is distinguishing between thinking this and the sentence in the book, so that the thinking does not have to be some talking or writing happening in his head (which of course could not be, but only be seeming to be). H says the thinking was quicker than the statement and goes on to talk of the nub of the statement, only fully articulated in the statement as such. This last bit seems wrong. The *non-linguistic* thinking (my expression of what he is after) is often much fuller, much richer than the attempt to linguistically formulate its content, although often more nebulous, even fleeting, there and gone. And H does say that the statement or utterance is indubitably a representation of some kind. Well does it represent the thought? Is that what is meant? Or is he saying that it represents some fact, namely the book being on its way to being finished? Or in what sense is my book a representation of my thinking? Well wouldn't we rather say that my book expresses what I think, is an expression of, directs you to what I think. Expression and representation are different. When you read my book do you have my thoughts or do you have your thoughts about my thoughts as expressed in my book. The word **representation** seems out of place, perhaps it enters because of other theoretical concerns or mistakes, namely mistakes about the theory of meaning.

Then *mental image* ('Ingrid sleeping on her side') and H saying '*what could the image be but a representation*'. But it is not that I see an image, the mental image is seeming to see Ingrid or whatever.

And then '*the outer sentence, so to speak improves a lot on the inner event*'. But how true is that. How does there being the sentence, my sentence, what I might utter or write, namely '*I need a cup of coffee*' improve on my wanting a cup of coffee. The sentence just expresses what I want and the sentence and the wanting are just as clear as each other.

From these various, recent considerations H *tentatively concludes* '*that our thinkings and wantings, since it evidently is the case that typically they are significantly **like** or anyway something like what indubitably are representations, **are** representations of a kind themselves, in whole or in part.*'

This seems to me completely wrong. And I think he is going on to say that these representations are **actual!!** But would I want to say using H's language that thinkings and wantings are actual but are not actual by being representations, which they are not. Well certainly they are real and are not to be dissolved in scientific metaphysics.

(And is it possible -this thought recently occurring to me- that H's talk of **actual consciousness** could be the same as contemplating an irreducible physical property. I.e. consciousness is actual, real, something in the world, something material. So not something that can be reduced to the language of physics and so disappear sort of, nor something existing in a dimension other than the physical world, e.g. a spiritual order. For instance H writes: ***The reason is that there is nothing abstract about our subject matter. As we are taking it, it has in it things as real as any others.*** Perhaps the point is that if consciousness is the datum, something we just find in the physical world like we might find sub-atomic particles and not reducible in any way, if you like, like the language of physics in its own right although not recognised as such, then the particular history of each instance, which we cling to as the essence of the matter -who we are individually- is like the elevation of a particular particle, because like anything else it will have history, when the levelling story is just that there are particles and consciousnesses.)

And then to note ***No doubt it is worth remarking in passing that none of this is to assume for a moment that language is necessary to consciousness.*** And H goes on to remind us that Descartes withheld consciousness from dogs and human infants because they lacked language, which H calls a 'startling conclusion' or perhaps he should have said 'the nonsense that philosophy can generate'.

H's treatment of *linguistic representations* is to suppose that from the science of linguistics we can differentiate and classify our linguistic resources (we might say here the elements of our thoughts, their building blocks) and these are *representations*, although H recognises a multiple ambiguity of the term (however, presses on with it). Then it is a matter of linking thinking and wanting to the building blocks, e.g. affirming *'basic propositions'*, which in turn H is prepared to regard as representations. This becomes the account of the cognitive and affective life of consciousness. So to be added to what he has said about perceptual consciousness. But this is 'clunky'. It is as though we have to have things that are at least quasi-linguistic for there to be thought and desire. Certainly in framing a linguistic account of consciousness we have no option but to resort to our linguistic resources, but such a concentration leaves no room for the reality of thought, desire and perception always coalescing and existing non-linguistically, and where the use of linguistic resources is just a means of indicating what has been thought, desired and seen. Again like consciousness itself we have to assert thought as the datum in its own right. We think the world not because of the languages we possess. In this respect there are affinities with the questioning of the linguistic fallacy in anthropology and elsewhere. E.g. an essay by **Mary Douglas** 1971 *'Do Dogs Laugh?'* (*'Speech has been overemphasised as the privileged means of human communication and the body neglected. It is time to rectify this neglect and to become aware of the body as the physical channel of meaning').* Perhaps then the challenge for computer science is to make a machine that thinks but which does not utilise linguistic or symbolic materials. Would that pre-empt the computer scientists starting point?

H continues his approach to affective consciousness by saying that an affective propositional attitude includes a *valuing*, a taking it as good or bad that something is true, a valuing of bare propositions. This is somehow prior to their external, linguistic formulation and this leads to H saying that *'we have a good deal more content for the proposition that thinkings and wantings are representations'* and then we have stuff like *'conscious representations are less articulated versions of linguistic ones'.*

It seems that H's actuality of perceptual consciousness is some version of the *unsynthesised manifold.* It is *what is there, a room out there,* and this is different from thinking of what is there and wanting what is there, these are, in contrast, representations, quasi linguistic or linguistic. But this is just following categories of traditional philosophy.

All of this seems a stultified treatment of desire and feeling. An awareness of a basic proposition and a valuing, a taking of it that it is good or bad. And it seems the propositional element, not fully and explicitly articulated linguistically is what makes for the representational element so that H can say thinkings and wantings *are* (his italics) representations. He passes by how what he is calling mental imagery and accompanying bodily facts, fear and rage, integrate with affective propositional attitudes (sort of the analysis of feeling and desire) as questions concerning finer points of phenomenology. And so a contrast builds for H between perceptual and cognitive and affective consciousness. This contrast is that of **actual** and **representational**. So it seems there are supposed to be these things, these subjective things, these actual things, this perceiving and then we start thinking and feeling about a subjective physical world. He is saying that perceptual consciousness is of e.g. a room out there, something external, whereas thinking of one's date of birth or what to do this afternoon are not actual as external. So I suppose he is thinking they must be representations if they are not actual!?

We then get the *Language of Thought, **LOT***. Signs, rules, conventions connected to what they are about. The language of all cognitive and affective consciousness. This now, for me, is seriously adrift. But **LOT** is to be distinguished from natural languages. It is deeper. This connects to Jerry Fodor, the Language of Thought Hypothesis **LOTH.** H seems sceptical of this particular offering and finds it difficult to see how it bears on the specific problems he is considering. As H investigates he does raise the question of language, in whatever form, and the thinking and wanting of human infants and dogs, who, and he affirms, think and want but are *unlikely* to possess anything analogous to human language. He asks *'Do they have a language of thought?'* Then he 'hedges' towards something... *'thinking and wanting are about things, stand for things ... That they are representative is* ***already*** *the fact that they consist in signs and symbols... that thinking and wanting is representative, surely, is* ***already*** *well on the way to taking it that they somehow involve a language somehow conceived.'* The idea here must be that thinking of something, wanting something is not in itself the actual thing thought of or wanted, and so, somehow, must be a representation of it. But this misses the point though, that thinking of x is a thinking *of x*, that wanting x a wanting of x, there is nothing here standing for. Like sensing, which is thinking, we think directly of things. Again the problem is one of not being able to bridge the gap. There is a gap but we bridge it, that is what sensing, thinking and wanting are they are bridges. We do not stand on one side of the chasm and only can picture what is on the other side, instead our resources of sense, thought and feeling enable us to span the chasm, and this is our being in the world. I'm inclined to widen this way of thinking by suggesting it goes with respect for **George Eliot**'s conviction that nature contains 'emergence', the principle of starting into life in previously unseen forms. A sort of genuine emergence of what's new; perhaps magic as a principle of nature.

And H falls into **the languages of thought**, as though thought is in a language, instead of it being that thought makes use of language, makes use of representation (not suggesting these are equivalences). He then sidles up to **Chomsky** and the science of linguistics, somehow the infrastructure of natural languages, underlying thinking and wanting. The language of thought at a deeper level. But then he starts talking about **leading the reader up the garden path** !!? So, *'But our subject is actual cognitive and affective consciousness, consciousness such that something is actual - representations.'* ... *'We come to the conclusion, then, quite against our initial assumption, that* ***LOT*** *does not at all consist in* *or contribute to an understanding of what is actual with cognitive and affective consciousness.'* This follows H saying that when he is thinking that *Kieran is a realist but generous* (one of his sons, I think) *'what is actual sure sure doesn't include a language'.*

H then quotes what is an often quoted passage in Fodor (in **Psychosemantics**) which supposes that physicists will in time complete the catalogue they have been compiling of **the ultimate and irreducible properties of things**. And then terminologically oblique to what I want to say denies this will be at the level of aboutness and intentionality. But putting to one side the narrowness of aboutness and intentionality my view is that sensing, thinking, desiring, feeling, intending, doing are ultimate and irreducible properties of things, and that physics is not the final arbiter of deepness. This is where everything has gone wrong, although there is nothing other than the deepness of physicality.

H clings to the question of consciousness and representation. We next get *evolutionary causalism, relationism and lingualism.*

'...all of us who are thinking about consciousness -scientists, philosophers, the lot- are trying to get ***beyond*** *the hold or grip and the common sense or everyday idea.'*

Evolutionary causalism seems a bizarre discussion. The idea that 'tiger', 'the frog pond' are representations and effects of causes, namely tigers and frog ponds, and that they persist in consciousness because their use has bestowed some evolutionary advantage, but that such theory does not distinguish between the evolutionary advantages of strong knees and representations, and so does not advance the investigation of cognitive and affective consciousness. It does not tell us what representations are.

Relationism is presented as a theory saying that *'cognitive and affective consciousness and the rest of mentality consists in representations that are physical, causal intermediaries that are also con-*

ventional semantic and syntactic signs and hence in connections of meaning.' This treats minds as **information processing systems** and so is **a computational theory of mind** and so a computational theory of consciousness. H talks of the pervasiveness of this science and philosophy of mentality, and how often objections to it have been put down by the professionals as *folk psychology* and *folk linguistics*. H takes a long time with this and I will not pursue apart from *'If'* a *'thought could be conscious and consist in only relational facts' a 'line of type'*, say this one, *'would be conscious. It isn't. Therefore relationism with respect to consciousness is false.'* And *'What of a sequence that turns up in a computer rather than on the page? There it is -a physical representation in virtue of semantic and syntactic conventions. It is no less silicon or whatever it is in virtue of its being a representation. You will anticipate that there seems to me no way of avoiding the conclusion that since we are at least in doubt about whether computer representations are conscious and they are* ***indubitably*** *conscious representations in cognitive and affective consciousness, the latter is more than the computer representations.'* Are we **in doubt?** I will only say/repeat there is no apriori reason why a machine cannot be conscious. In fact it is an inference from consciousness being a physical property that a physical thing like a machine could be conscious. This though is different from allowing that representations could be or are conscious. It seems more natural to talk of being conscious of a representation. The linguistic form of words in my mind (some p, say) I will be conscious of, and conscious that this linguistic form is in my mind, but p is not conscious (i.e that H's wife is in the greenhouse). And if there are physical analogues of p in the brain, when p is in my mind, they, just like the physical sequences in silicon form, will not be conscious. The only way for relationism to go is to deny there is anything that is consciousness. And how absurd is that? And is not this concentration on representation misguided. The thought, the desire, is of x, say H's wife in a greenhouse, and then there is this temptation to say that somehow the thought is a representation and H's wife being what the thought represents. But this is because we cannot understand what thought is. I argue the thought of x is not a linguistic proposition that may express the thought (that is the role of the proposition in thought), nor is the linguistic proposition itself a representation. 'H's wife is in the greenhouse' does not represent whatever it is the thought is of, or not in the sense of being a representation of it. The point is that the proposition is not like, in any relevant sense, H's wife in the greenhouse. In a weaker sense we might say that **p** can stand for and in that sense 'represent' the fact. Analogously we might make a stand up **HW** (i.e. the letters on a base) and then on a table, say, place a box, and place **HW** in the box, then say that, pointing to it, 'this stands for H's wife in the greenhouse' or 'this represents H's wife in the greenhouse'. We would say such things. This seems acceptable thinking of demonstrations, but how strained is it to say of a novel that its sentences stand for or represent what the novel narrates. A lot of care is required here not to accept a few limited examples and suppose the philosophical assertions fit in with the complexity and anti-essentialism of ordinary language. But returning to H's thought and possible desire, what **p** expresses H **thinking** of his wife in the greenhouse? Certainly not 'my wife is in the greenhouse'. If H is thinking of his wife being in the greenhouse, ie. thinking that she is there as he thinks and more positively thinks of her, then **p** ('my wife is in the greenhouse' or whatever) is an entailment of what he thinks, but his thinking of his wife, what this is, is not **p**. H's thinking of his wife is not a proposition, it is thought.

And because it fits in my envisaged project as a whole I quote H on **Turing**, without knowing, in advance, where this fits in with H. Speaking of the lack of care in designating the subject matter H says '*Was there that lack, despite his degree of care, in the seminal work of the admirable* ***Turing (Hodges)*** *as well as many who have been attracted to the* ***Turing*** *Test? (****Oppy*** *and* ***Dowe****). There is or can be a lot of looseness in speaking of a test establishing that something* ***thinks****, or* ***has a mind****, is* ***intelligent****, or is* ***like us****, or* ***passes as an imitation of us****.*

Lingualism (H still considering other theory on his way to what is promised; a clearer view of actualism). Stalking horse Searle. What is worked hard here is a relation between conscious and linguistic representation, where the properties of linguistic representation, which are specified, are modelled on those of conscious representations which are not specified, except by analogy with linguistic representations. H hounds this for a coherent account of the difference but fails to find it, and so finds the position insufficient, inadequate. (For Searle, apparently, aches and depressions are not representations. Is this because they are what they are and cannot plausibly be held to mirror something else. Seems a strange pairing anyway!)

That thought is understood as conscious representation (basic propositional content plus an attitude towards it) is, **though, wrong from the start.** What is the attraction of *representation*? The idea that thought is somehow a mirroring, a correspondence, as though thinking is a sort of image projected onto a screen, and that language mirrors this and in itself mirrors. But consider, *seeing* is thinking and thinking is a process, so that trying to account for consciousness through the concepts of a thought and a proposition must be stultifying. In *seeing* there is no mirroring, there is nothing that corresponds, and *seeing* is, if the terms have any application, both cognitive and affective consciousness, and not some sideshow called perceptual consciousness. Looking and seeing are thinking and there is nothing apart from that seen and the seeing of it and this is the object thought of and the thinking of it. Mediation is not required. There is no representation unless hopelessly we start to chase the image on the retina, and clearly this is not the way to go. But what of imagination which is also thinking? Does this not require some notion of representation? But again there is nothing, and certainly nothing that shares some affinity with what in this theory (lingualism) is called linguistic representation. Instead, in a basic case, we imagine something, which is thinking of that something, which we might say is like seeming to see that something, but the seeming to see is not that of looking at a picture: there is nothing to look at. Representation is an unnecessary element, which intrudes because we cannot grasp what thinking is. Propositions issue from thinking but that the basic structure of thinking is propositional is like trying to put thinking into a **straight**-jacket; failing to grasp the fluidity and incoherence of thought, its unresolved nature (intrinsically). Thinking is never finalised but how it issues in language sketches thinking. Language is a working hypothesis, or rather, its uses are.

Leading up to H's **Conclusion Actual** we do get an elaboration of what conscious representation is and conscious representation, in its turn, seems some clarification of what thought is. Conscious representation is said to mean something and is said to have something of what it means. The exemplification of this is, what the conscious representation means, it e.g. (from a long list) pictures, portrays, corresponds to, is a sign or symbol for/of, I suppose the havings of something that it means. I have already cast doubt on this as an approach to thought.

We then in **Conclusion Actual** get H's departure from conscious representation as a dyadic relation, so '*Thinking and wanting*' (cognitive and affective consciousness) '*somehow include three facts - a representation, the represented thing, and the fact of representation somehow being actual.*' Reintroducing the notion for H's exposition of perceptual consciousness. A clue to the meaning here might be in the quotation from **Strawson** of which H approves, '*Consciousness is, indeed, that which alone makes all representations to be thoughts...*'. Further clarification. 1. For cognitive and affective consciousness there is no counterpart to what is actual with perceptual consciousness, i.e. a subjective physical world. 2. Perceptual consciousness bereft of representations but not so cognitive and affective consciousness. Then we get the distinction between cognitive, affective consciousness on the one hand and perceptual consciousness on the other has to be made because *'Is it possible to forget what it is to close your eyes and think of England'.* And so what is supposed to jump up from this is a representation (an actual one!?). But this is too hasty and anyway neglects how cognitive and affective consciousness are always a part of perceptual consciousness, or rather these distinctions are distortions of consciousness. And then the actual bit is perhaps clarified in *'What was actual included such a reality -an orientation or relation to a bare proposition.'* (political passion, occasional rage, moral rage) Is this the actual bit? But is the attitude a representation? H raises this as a question. Or could it be a bodily manifestation? And then *'a representation being actual ... includes not only the thing or property that means but also the thing or property that is meant.'* All of this discussed with reference to a range of examples, like Highgate Wood, the gate to Highgate Wood, the King of France, the unicorn etc.

(There is a review by ***Dale Jaquette*** *Universität Bern in Notre Dame Philosophical Reviews, electronic journal, 2015, showing I'm not alone in finding it difficult to interpret H.* ***'I am not sure that I fully grasp Honderich's distinction between objective and subjective physical reality that is key to understanding his new theory of consciousness.' 'To the extent that I understand the concept, each of us lives, functions or operates within his or her own subjective physical world.'*** *DJ concentrates on perceptual consciousness claiming H gives most time to this, which I think is not true. He raises questions about how subjective physical worlds, experienced individually cohere with an objective physical order -giving a Strawsonian/Kantian slant to the latter-. On this approach we are not far from all the problems with solipsism. Interestingly for my purposes he does compare H's narrative, such as it is, with* ***Tristram Shandy*** *- H is said to drift- and this is, although he does not say so, how consciousness is. There is not much to this short review but its main thrust is to question the tripartite distinction of perceptual, cognitive and affective consciousness as definitive, pressing instead for*

some transcendental consciousness which is aware of all the other awareness. I don't think this is penetrative.)

Perhaps what H is after is that cognitive and affective consciousness is a representational mode, which is not just what refers and what is referred to, not just an aboutness and what the aboutness is about, but is in addition a psychological reality, perhaps even, something happening in the body, but not an externalism, not a subjective physical world (the latter required for H's exposition of perceptual consciousness). And what do I disagree with here? Well to the fore it must be this concentration on representational mode. When I think of the **King of France** there is no representation. There is consciousness and this is thought (**Strawson?**). Thought is a physical event in the body not to be equated with neural events: correlated but not equated, different facets of the physical. The understanding of this is consciousness.

H reaffirms: '***A thought or a want is not identical with a line of type or some silicon. Conscious representation is a matter of both lingual representation and actuality.***'

And '***Conscious representations are like linguistic representations in presupposing languages. Conscious representations share a character with linguistic representation - lingual representation.***' But '***Conscious representations are less full and explicit than their counterpart linguistic representations.***'

Then H gives us a very limited approach to the lingual aspect which he seems to think a necessary part of a thought/conscious representation (a necessary part but requiring **actuality** to turn lingual representation into conscious representation). He raises the question of how we are to think of what he calls the conventions that are integral to representations. And elucidates this by pointing to the way a child is given a name, which is an absurd narrative on his part in which parents agree to give a name because mothers require a way of discussing the offspring on the phone with others. Something or other is chosen which then stands for the child and this is taken as exposition of the elements in lingual representation and representation more generally, i.e. representation is established by means of agreed conventions. All of this preceded by the question of how the conventions came into being. The answer seeming to be the selection of a sign (arbitrary?) to solve a pragmatic issue. But if we are asking about how things **came into being** there would seem to be a prior matter which is before a dictionary of signs, namely how is thought explained. Is it possible we need to drop representation? This is what I argue. Absurdity with H continues as he moves on to discuss (briefly|) absent denotation (unicorn, **Pegasus**). I suppose for H the question seems to be about how the sign convention is possible when there is nothing for the sign to stand for. His very brief answer is to gesture towards the empiricist-creation theory, via **Locke**, **Russell** etc. Hopeless!

H then revisits **Searle**'s Chinese Room Thought experiment. The point of the experiment is to argue that we need both semantic and syntactic rules for there to be understanding. (Chinese symbols not understood, someone who does not understand Chinese but following rules to provide outputs for inputs, i.e. providing the symbols required by the inputs without any understanding of the meaning of either the inputs or the outputs. Just a matter of following rules. So satisfying the **Turing** test on performance. The inputs are questions, not understood as such, and the outputs are all answer which are correct. Performance results are indistinguishable from those of a competent individual fluent in Chinese. But the missing element is using semantic rules and so understanding meaning, and so understanding. H uses this approvingly but wants to add actuality to semantic rules to give a full analysis of thought. **Searle**'s distinction is not enough. (Now there have been lots of objections to **Searle**'s position and I suppose some of them might point to how our computer systems having the resources of being able to translate the signs they manipulate, but that is not really to the point. The problem is that Searle can't get the experiment off the ground without having an agent who in the first place understands. What is there that is equivalent to this when a computer performs its tasks? That is in what way are these situations equivalent? It is as though **Searle** is so impressed with the similarities at the back end of his thought experiment (i.e. Chinese symbols not semantically understood but manipulated according to rule bound procedure) that he fails to attend to the front end. The human participant at the front end is full of understanding but the computer? We are in the dark. Does it do what it has to do in the same way as it can manipulate the symbols that it syntactically relates but not semantically? Or could it understand the task, i.e. not just perform it? I can't see why H is so impressed with **Searle** here. It is superficially clever but the point it makes is seriously underdeveloped.

H then suggests that maybe it is not that actuality has to be added to syntactic and semantic conventions in order to get to a non-circular treatment of conscious representation, but that actuality is to be found within semantic and for that matter syntactic conventions, leading, perhaps, to a proper understanding of them. There may be something promising here.

Then **10** with a sub-heading, namely, **Being Actual Is Being Differently Subjectively Physical**. This section opens, as follows:- '*We have asked and answered the question of what is actual with cognitive and affective consciousness. The very short answer is representations.*' But H asks what is it for them to be actual? The answer is that they are subjectively physical, both physical and subjective. Actuality cannot be the represented thing existing, because of unicorns, the king of France etc. A first try at the implications of 'physical' is to remind that dictionaries differentiate between body and mind and that which is demonstrably physical is that which science he starts to make a distinction between contemporary and future science) accepts as physical properties. I suspect the introduction of future science is to allow that representations will in the future be classified/ count/be seen as physical and quite correctly, at the same time H introduces physical, neural events as being alongside conscious representation and even accepts as proven that neural events are temporally prior to those conscious representations associated with them. But then the conscious representation will be antecedent to any conscious representation and prior neural event which it leads to. Thus even if there is some neural event prior to thinking what is the sum of 7 + 5 nonetheless thinking what is the sum of 7+5 will be antecedent to thinking the answer 12 and thus antecedent to any prior neural event before thinking 12.

Exploring the physicality of consciousness H directs attention to how it is the subject of scientific inquiry, e.g. how doctors ask questions of our pains and how optometrists ask questions of what we see. Straightforward applications of scientific method in an applied context. In H's view it cannot be that if there are things which cannot be measured that they are excluded from science. Then I think he shows that he is not arguing what I am arguing. He says when he thinks of his mother, i.e. when he thinks, as he puts it, of Rae Laura Armstrong Honderich, there is a representation whether by her name or 'that old image of mine'. So it seems thinking is necessarily lingual or pictorial, what he wants to call a representation. I am saying something quite different i.e. that he can and does think of his mother but without some silent saying of the name or seeming to see her. H goes on to say quite rightly that the lingual or pictorial representations are not in space. Although, and he is not saying this, my thinking of x is in space because I am a thinking **thing** and so necessarily my thinking takes place in space. The trouble with representation is that that there is no such thing and so cannot exist in space. H says '***It*** (representation) ***isn't somehow existing but nowhere any way. Rather than being nowhere, I guess I might be tempted to say it's* around here somewhere.**'. However he finds no difficulty in attributing some temporal measure to his thought, but then asks if something can take up time but not take up space (what he calls a deep question). This then is H's rather painful wrestling with how conscious representation shapes up as the subject of a scientific investigation. H then goes on to say that conscious representations are in space as anyone who has had a decent education knows and as any neuroscientist knows because thinking and wanting are **events** and how can there be any events that are not in space? But this is hopeless confusion! Thinking and wanting are not all the same as conscious representation. Of course though I agree 'thinkings' and 'wantings' are events and so are in space.

'**We know where a representation is - in a head**' p315. So it has spatial properties.

(And an insert from the review of ***Susan Greenfield****'s* ***A Day in the Life of a Brain*** *by* ***Adrian Woolfson*** *TLS Dec 9 2016. Exploring the project of translating from brain states to conscious states Greenfield directs attention to 'neuronal assemblies', i.e. transient collections of large numbers of interconnected neurons, which have proven amenable to mapping from the 1990s using VSDI and fEITER techniques (what these are need not detain, but there are such techniques although VSDI cannot be used on humans!! Woolfson says that Greenfield's book does not address at any point how a neuronal assembly can generate a conscious event. There is ambiguity here. How x can generate y and how x can be translated to read y are very different. My view is that translation has to be asymmetric from y to x. With 'how does x generate y?' I'm wondering if it is a question. Is it like asking 'how does x have its properties?' H has mentioned Greenfield and 'the head'.)*

'Think of representations close together in time. They are at least typically causes and effects of one another. Mother brings Father to mind.' p316 *!!!*

H goes on to make causality, i.e. lawful connectedness, the criterion of physicality and then wants to say that conscious representations are causally connected with behaviour.

'We aren't epiphenomenalist because we take conscious states to be certain neural states that do cause actions.' p316 This is H's refutation of dualism. 'How could what is unreal … bring about what is real?' And all this emphasis on causality and physicality leads on to the doctrine of empiricism, so that there is externality and subjective physical worlds and conscious representations all linked together in a causal order.

H then raises the spectre of determinism. Is what he is saying embracing determinism? The positions seems to be that determinism probably is one of the better options but that the theory of consciousness H puts forward is not dependent on it.?

(Perhaps the best reading of **actual** is physical. Consciousness is physical. There is this interconnected causal order which confers physicality and actuality on its constituents. There is a physical externality which is the cause of a subjective physical world -the world of perception- which in its turn causes conscious representations (the empiricist story) occurring in physical space, which in turn cause physical movements (action), which cause changes in the subjective physical world and so changes to externality. All of this actual, real, physical, causal and for H law-like and so possibly determined. And remember H allows changing viewpoints on the subjective physical world but not on conscious representations. **SO, perhaps there is some convergence between what I am saying and what H is saying, leaving aside causality and determinism. Thus I am arguing that the world we experience is a multiplicity of appearances (the hillside can be both red and green, which we see being depended on our optical constitutions). Is this not in some sense a subjective, physical world. What is not given directly to sense is the atomic and subatomic substance of what's physical. I do not want to divide this distinction into different orders of reality, namely subjective, physical world and its ex-**

ternality. I am saying that the multiplicity of appearances are just properties of the physical. (In what sense is what a thing is the cause of its properties and so if you can't say that you can't say that externality is the cause of a subjective physical world, if this latter concept is understood as I am suggesting it might be? To divide things up as H is suggesting is to simply duplicate traditional and incorrect philosophical dimensions of reality and then to hang a tag of physicality onto all of them but otherwise leaving them as they are.)

'... what is actual with perceptual consciousness is different in kind from what is actual with cognitive and affective consciousness, and also their actuality or physicality is different.' p 319.

H then, concentrating on 'representations', invokes privacy as a property of them, but only contingently. He thinks this could change in the future and imports the concept of science fiction to point the way. He also invokes the concept of self-deception with regard to what he is calling representations, in the sense that you can be self-deceived about what you think or want and others may be in a better position to know what you think and what you want. And then there is confusion and forgetting. **Do I have anything to say about what we might call misfirings in consciousness? And how are they possible?** H then goes on to talk of a **unity,** which seems to be a bringing together of all the bits from what I have been complaining about, namely an atomistic analysis based very much on traditional philosophy. He is talking of subjective, private or personal unities. Individualities. And then, from **Searle, networks** and **backgrounds.** I suppose bringing all the atoms together as part of a functioning machine! Then, **Kant**, and the Unity of Consciousness. A lot of jargon here, e.g. synchronic unity (thinking of **Kant**, the room in which you think, and the feeling of the chair on which you sit in the room and think) diachronic, e.g. things coming together from moments in the past with present thoughts.

Next to the comparison of subjective physicality of cognitive and affective consciousness with the physicality of perceptual consciousness. (p. 322) **And the physicality seems to depend for H on actual consciousness coming within the inventory of science and being open to scientific method.** So he goes on to construct a checklist of their characteristics. In addition to being open to scientific scrutiny subjective representations occupy space and time although H thinks some uncertainty as to their spatial location?! **I suppose he means 'exact'. Is he talking about spatial location in the brain!?**

Lawfully connected to other representations of (**and here we get the word**) the 'representer' and to other categories of things. **So this is some gesture towards causal relations and possible determinisms**.

But representations '***are not physical in the sense of being perceived in the ordinary sense .... But do have dependencies on subjective physical worlds'*** (perception)

They do not have primary and secondary properties and do not have points of view or are different from different points of view, and H says not 'even images'.

This is the end of the list of their supposed **physicality** and H then proceeds to consider their subjectivity.

Actual representations cannot be separated from consciousness

They belong to one particular representer. This representer does have some privileged access to them.

And then we get the stuff about self-deception and truth and remarks on how representations integrate into unities!

This is some of what is meant in specifying these actualities as physical. '***A conscious representation is a lingual representation that is also actual ...***'

We then get, with an excursion into Fodor, that thinkings and wantings are physical but different from what H says is called the **irreducible physical**, the deep physical i.e. what physics understands by the physical. And here is the mistake. The uncritical respect for physics, as though it and scientific method and lawfull dependencies are something of a higher, unquestioned order. Ultimately this is acquiescence to the demands of economic reification. And you can see H playing around with the possible implication of the non-existence of representations, only to deny this both for Fodor and himself. So perhaps physical but not deeply physical. I want consciousness at the same level but because there, physical and existing, changing everything. H wants to talk about '*subjective physicality*' that's what the representations of cognitive and affective consciousness consist in. Then H says these things are not less physical than the objective physical world. (But that is because they are part of the objective physical world!)

And this might be it as we go to **11 Conclusions Past & Present**

'You can proceed further by considering what it is for something to be physical in at least the dominant sense in science and philosophy. That is, what is it to be **objectively** ***physical.'***

'..universal representationism is denied by actualism.' so rooms exist and are not representations of anything. This is H's beef about perceptual consciousness. The actual is **subjectively physical**. ***'One physical characteristic is spatiality. One subjective characteristic is ..... Individuality.'*** And then '***Cognitive and affective consciousness is actual representations, all about or in some derived from subjective physical worlds or thereby from the objective physical world. ...there is the objective physical and there is the other physical.' '...consciousness is a reality itself in space and time and that it has physical effects ...'***
'Consciousness is real ...... what is real takes up space and time, has effects, and so on. More generally, it is physical.' 'Actualism makes consciousness real .... by making it physical.' '... a good deal of philosophy on consciousness does not concern itself with the

matter of physicality .... Is there any philosophy on consciousness that has supposed that progress on its very nature lies in getting clearer about physicality? We seem to have shown otherwise.' Subjective physical world: ***'a world out there in space ... but not independent of the perceiver. My subjective world now has a dependency on me .... That unconscious mentality does of course include kinds of unactual representations and so much of the rest of irresistible science, including the science of vision.'***

'perceptual consciousness - ... the fact of your contributing to what you see doesn't move what is outside of you inside you, doesn't turn it into something neural or whatever. Or, more carefully, the fact that your seeing something right now is something outside of you being actual, and that your having retinas and a visual cortex, doesn't move it an inch towards inside you.' p 341

Personal identity and what H is calling individuality are then dismissed as being some mysterious self (that's no surprise) and instead treated as the being able to link through personal memory the experiences of one's life (**Locke**).

H on what he is calling naturalism: ***'consciousness and all the mental are "elements of the natural world, to be studies by ordinary methods of empirical inquiry".*** Chomsky **quote.** And then H goes on to say ***the facts of consciousness are those of one of the two domains of the physical, the subjectively physical, the other domain being the objective physical.***

(The concentration in H on perceptual, cognitive and affective consciousness preloads things towards how consciousness is with the one who is conscious. Thus what is it like to be conscious. And this explains so much of the presentation whereby the reader is addressed to compare being conscious with the account being given. So 'dear reader forgive me for presuming but I put it to you that this is how it is with you, how you live your life.' Not an actual quote but there is much of this. **However** the perspective we do not get is the obvious empirical data of beings who are clearly conscious. I.e. we observe consciousness all around us, the world is a world of conscious beings, part of plain, ordinary experience. We see consciousness. Perhaps the nearest H comes to this is, after setting up what I will call (without much care) an ego-centric account of consciousness, is to hand over this subject matter to science, but that is only a small part of a social perspective.)

There is a book **Mind and Consciousness: Five Questions (Grim)** where 20 philosophers are asked whether a science of consciousness is possible and 10 say yes including three objective physicalists who say the subject already exists.!

For H ***machine consciousness*** is a possibility but remains science fiction. (I suppose another way of presenting my own project is to say that I try to specify what is required for machine consciousness to be a reality. And to resist the concept of animals being just **conscious robots**)

H says he hopes it is a misunderstanding of the Turing Test to say that ***anything indistinguishable from a person in answering questions is*** **ordinarily conscious**. What he thinks is a proper understanding is to say it has unconscious mentality (intelligence, calculating ability, etc) and for H something which we share with these machines, part of a causal story.

The question of how consciousness is generated by the brain H calls the mind-body or mind-brain question or t**he hard problem**. H says it is pessimistic to argue that there is something mysterious or unintelligible about the connection between brain and consciousness. The criticism of pessimism for H must mean an optimism about what science will achieve and that the pessimism removes the subject from rational investigation. This is something my account faces with its notion of consciousness as irreducible, physical property.

H says what physicalism has to do is to get rid of correlations between the neural and the non-physical. This, H, along with **Searle**, **Dennett** and **Papineau**, does but with H's physicalism, subjective physicalism, he preserves a difference in kind between consciousness and correlates - ***physicality with real difference.***

There follows a section called **Pessimism? P 348** This makes me feel part of the flow. It is at this point I invoke a concept reworking what could be meant by **magic**. H says, ***Some mysteries concern mind, and they may be insoluble forever.*** Nagel and Chomsky we are told hold that there are mysteries, something more than problems. Mcgin ***mysterianism. 'A miracle beyond our science'*** (Of course I am arguing that this in some ways is not mysterious at all because the world has to be a certain way without explanation, that is its explanation.) Why the experience occurred? In some sense lawful correlates don't help they are just brute correlations. H plays around with intermediate steps in a causal chain still to be discovered and also with the notion of conceptual in addition to lawful connection. H is after how actualism clarifies the issues of mystery. But what we get is that it offers something quite other than lawful connection, a strict identity and X=Y, thus ***'your being perceptually conscious right now is a room out there, a room's being subjectively physical'***. This a conceptual connection!! (This too brief and I've been following for 350 pages. Could its alternative to mysterianism just be, this is the physicality of the world? And we do get p354 ***'Put up with reality as it is.'***)

H sets up the question ***'Is actualism really identical with naive realism?'*** But for H they are unlike because naive realism doesn't specify what it takes consciousness to be. (But this may not be to the point because direct perception is just part of the exposition of consciousness not a theory of consciousness in its own right.) ***'It is not a theory of all of consciousness'*** (Well that's my point. Why is H so worried about competitors?) Yes there is a difference and H spots it, namely, naive realism is not saying that there is a subjective physical world. But then SPW is Honderich's mistake.! And remember right at the end H refers to **P F Strawson** and common sense realism in his paper **Perception and its Objects.** And in conclusion there is a skirmish with determinism and freedom and H supposing that what he is calling **origination** wants us to be over and above the rest of what exists i.e above nature (But this is all wrong. We have to redraft our concepts of nature and physicality that's the problem.) H's position seems to be that actualism puts humans and maybe some others at least in a differentiated position compared with the rest of nature but that this is in no way a defence of origination, which is just dreaming!

(The concluding section of the book is endlessly repetitive. Constant listing of the rejected theories of consciousness, constant listing of the bulwarks of actualism. And a lack of clarity. Throughout the book there is :- Intellectual snobbery (scientific attitude and philosophical logic). Patronising references to other thinkers as though terrified of rivals (the vicious competitiveness of academic philosophy as remarked on by feminist philosophers, US competitive capitalism and its promotion of analytical, non-political philosophy).

A little clue towards the end, '***No one can think of excluding the other category either, say, the unconscious events that take place between my trying at a moment to remember some mere politician's name and then my doing so.***' H wants a lot of this to be part of the causal story explaining behaviour. The trouble is does he think that the unconscious events are like conscious events but unconscious, e.g. some representation? And H goes on to talk of specifying sufficient conditions for behaviour, causal circumstances! The dismissed concept of origination might suit better.

And towards the end we get '***consciousness evolved***' and '***That actual consciousness conferred survival value on what are loosely called competitors is as evidenced as anything in the world.***'

ALAN TURING. (plus wilder speculations)

In promoting non-mechanistic physicalism should some consideration be given to the foundations of IT and the possibility of a wrong turning taken at that stage determining modern culture?

In the late 1940's **Turing** discussed brains and machines (not minds/ machines nor consciousness/machines) and queried the idea that the 'the two things' are separate by challenging those who thought they were separate to devise an ***examination paper*** that could be passed by a 'man' but not by a machine, indicating a reflex to suppose what what makes 'man' 'man' is an intellectual function or perhaps better an intelligent function, so that what counts as intellectual is logical, mathematical problem-solving which, he envisages, machines could do as well if not better. The implication then is that 'man' is just a special sort of machine. But then just as **Turing** finds himself in this position his commitment to scientific method makes him review the possibility of counter-examples and he, for a moment, thinks a testing counter-example is the unpredictability of human behaviour, but his immediate reflex to support his questioning of the main distinction is to counter by the thought of installing a roulette wheel in the machine! But this shows no real grasp of what consciousness is. The point is that the reflex is one of simulation not of identity. At the same time **Turing** wants to talk of machines thinking, 'the machine thinks', 'machines can think'. The practical application of this leads to work on chess-playing machines, as though this might have something to do with the matter. The supposition must be that chess-playing somehow is the paradigm of thinking and achieving the goal of machine-supremacy. But what is not considered here is a dialectical reversal. Namely, that consciousness simulates computability succumbing to its own reification, as when children learn their times tables. Consciousness has the ability to mechanise its own abilities, without ever becoming a computer. **Turing**'s view is that the machine simulates intelligent functioning up to the point where it becomes indistinguishable and then surpasses. Two machines, one in the end demonstrating its superiority. But computational consciousness is consciousness simulating mechanistic functioning. This we may call intelligence but is not at all the same as intellectual functionality. We can turn ourselves in part into machines and this need not be confined to the arithmetical and the algorithm, as, for example, when monks resort to repetitive chanting to reach what they may call nirvana; a state of reification, not that they would describe it as such. It is true that **Turing** in his 1949 **Mind** article **Computing machinery and Intelligence**, where it is his view that machines could imitate all human response so as to deceive any interrogator, nonetheless concedes that there is a 'mystery about consciousness', but this is because he cannot fathom its material base. However the point is, as I have argued throughout this book, consciousness is not a mystery but instead something self-evident and self-explanatory, and only looks mysterious if one is seeking elusive, mechanistic, reductive causation. The notion there will be in time a natural translation from neuronal spike trains to a language of consciousness, an uncovering of equivalence, just leaves a problem as large and insoluble as the Entscheidungs problem, only in this case an unnecessary problem.

To enter the complexities of what needs to be fathomed compare a Turing Machine with a knitter and a knitting machine. The knitting pattern is the tape, the rules, the instructions and knitter and knitting machine are driven by the pattern. Knitting is then robotic. But following the pattern for knitter and knitting machine is different. We would say the knitter sees that it is so, that for the next four rows it is 4 plain, 2 pearl, the machine just makes four rows of 4 plain, 2 pearl. The process of making is dissimilar. What is universal is the programme. However, the knitting machine can mal-function and produce an output not prescribed by the pattern. Similarly, a knitter can make mistakes, knit pearl when plain is prescribed, drop a stitch, gain a stitch etc. But also the knitter may make judgements, that is *see* that the garment is not going to work out. The knitter may have spotted an error in the pattern and then adjust to produce an acceptable garment. The knitting machine, on the other hand, would just go on following the instructions, go on being driven by the instructions. The difference is seeing that it is wrong. This is the accompaniment of consciousness which distinguishes knitter from knitting machine. This would be true of standard knitters and knitting machines. A knitting machine programmer might add sensors that would react to infringements of dimension requirements and so trigger a sub-programme to determine a different output. This would be simulating the knitters seeing an error. Turing is right to question whether the knitter can arrive at an output which the knitting machine cannot. So we might say any output can be achieved by the knitting machine/Turing Machine and this we may not be able to say of all knitters.

What is fundamental is that seeing that x is the case is no more and no less than that, and for the subject self-explanatory. As an accompaniment to process it will itself be accompanied. Accompanied by physical events. These physical events are conscious and there would not be consciousness without these physical events, but consciousness is a self-evident, self explanatory property of the physical event. The physical event is not the means to 'seeing that x', 'seeing that x' is a property of the event. In the old way of saying things it is a primary quality: it makes up what the physical event is. Perhaps in Turing's way of saying things it is an *intuition.* And we all understand this until we look to simulate this, when we convince ourselves that the simulation is what is simulated. Consciousness as a real, physical property has all the implications of the language of consciousness. It is, as what it is, a causal property in its own right. It is a 'free radical', exerting upwards, downwards and sideways causation, and so capable, as Marx says, of changing the world.

So the **Turing** prediction/thought was that there would come a machine such that we couldn't tell that it was a machine, and so then everything would be a machine. But this isn't the point. The point is how do we tell the machine is conscious or not? What consciousness is, is irreducible and so not open to reductionist analysis. This is not the same as saying consciousness is unanalysable. Fundamental particles have consciousness as their property. In a way what's at issue is obvious. A machine following an algorithm arrives at a conclusion, an output, identical to that reached by conscious 'intuition'. It may arrive earlier or slower but the fundamental difference is between the ways the conclusion is reached. Computer science need have nothing to do with creating consciousness, but it misunderstands itself if it supposes that identity of output achieves this. It is like saying that **Leonardo**'s **Mona Lisa** is Mona Lisa. A problem with AI and simulation is that what is simulated is rationality par excellence. The computing machine is never unintelligent. It relentlessly executes its algorithms and if there are mistakes or errors they are not attributable to its dumbness because computer machines do not think, they simulate a particular, meritocratic model of thinking. This is not to say that in the future there may not be a way of adding consciousness as a property to a computer machine.

At the limits science accepts something out of nothing (Big Bang) and something out of nothing is an ordinary principle of existence. Matter is creative. What fills space and time is creative.

If consciousness is an irreducible property of what is material, when present does it add materially to what is material? E.g. does conscious matter weigh more than matter without consciousness? Clearly there will be a difference in electrical activity. Does a light weigh more when switched on? Here *weight* may be a dead herring and I may be out of my depth. But to a totally wild surmise:- if for there to be matter there has to be anti-matter, could consciousness be anti-matter? Is what fills space and time conscious? Just as a human body has consciousness as an irreducible property so can we surmise that all that fills all space and time has consciousness as an irreducible property and this is !!! anti-matter. In the offing would then be questions about gods and pantheism, but dismissable if consciousness is a property and so could not create what fills space and time.

Playing chess is constrained thinking. It is forcing thought into algorithmic mode. This is how we can design machines with this mechanical function. Consciousness has discovered this mode of solving problems, but thinking is not like this, although it adapts to do so. **Turing** thinks of thinking as though it exists to be examined. But thinking is something else and thinking uses computing machines to free itself from algorithmic repetition. Without a question, without a task, without a programme, a machine does not even know that it doesn't know what to do with itself, that is because it is not an itself with something or nothing to do. What has to be grasped is that AI outstrips human computation but this is not proof of mind. In computational tasks human consciousness or conscious imagination simulates being a machine. AI dehumanises consciousness. Chess players, mathematicians turn themselves into machines. Some conscious thinking things are better at this than others, i.e. better at alienating themselves.

Turing's thinking is the product of algorithmic, social conditioning. A system of pressure and constant examination. A social imperative of developing a technocratic function. Turing is programmed by an educational system seeking meritocratic reification. The production of nerds, socially required by competitive economy. Sexuality is his deviation from economic conditioning, something completely at variance from the stringencies of logical straightjackets and so indicative of the independence of consciousness. The Turing Machine is seen as the world solved. The followers of **Turing** claim as much: 'the most important single intellectual development of the C20th, and possibly of all modern human history' (**Wolfram**). **Turing** seems to be the **Pythagorean** ambition fulfilled, a universal mathematical logic able to represent and so solve everything, the great digital reduction, the structure of everything, and most significantly, mechanistically, electronically enhanced, the fusion of a physical machine and a virtual machine. This leads to the view that **Turing** is a mechanistic, behaviourist theorist of the mind, intent on reducing the concept of meaning to that of information. Perhaps though a characterisation not altogether fair because although addressing greater formalisation his approach is still pragmatic, still open to opportunistic pluralism and usefulness of communication in the design of notation. Despite this it is the measure between **Turing** and **Wittgenstein** that defines the project. **Turing** is not a humanistic philosopher of meaning and 'forms of life', hostile to mathematical logic and the very idea of a **Turing** machine. The one minor skirmish between the two is indicative of diametrically opposed views as to the structure of reality, an issue of the relation between formal contradiction and reality. **Turing** suggested that formal contradiction might mean a flaw in the world so that a bridge might collapse but Wittgenstein says that it did not seem **quite right** to say a bridge might fall down because of a contradiction in formal logic or mathematics. To extrapolate this is the beginnings of metaphysical intrusion into computer science. The logical, ontological divide.

POSTSCRIPT

The indeterminacy of consciousness and its escape from mechanism but not physicality can be compared to what would be regarded as the real science of the 20th century including e.g. quantum Mechanics, quantum entanglements, Heisenberg**'s indeterminacy principle, the** Einstein/Bohrs **dispute: theoretical contradictions at the heart of mathematics, which have led classical physics away from what it regarded as common sense positions, although positions far removed from naive realism. In place of the classical view, theory has been prepared to postulate instead the 'spooky' action of indeterminate particles upon each other; indeterminate unpredictable entanglements, and so, for example, to Google the topic (after all a main conduit for information) a concept of magic does begin to enter the conversation echoing the concept of 'spookiness' and leading to uncertainties like** 'it's not magic , it's quantum mechanics'**: in other words if not asserting something, establishing an association. And so despite the quarrel with science throughout this book a certain affinity arises between it and consciousness and an affinity which if reflected on perhaps begins to explain the paradox of quantum theory, allowing a mirroring between mathematical reality and irreducible physical consciousness and so not completely sidelining the claims of mathematics to grasp the material world. All of this interpreted in this way challenges a theory of continuants.**

NAME INDEX

(This book is not scholarly, it is a broadly conceived argument referencing a few sources.)

R. Descartes
Charles Dodgson
Mary Douglas
Phil Dowe
George Eliot
Albert Einstein
Michel Foucault
Jerry Fodor
Alan Ginsberg
Susan Greenfield
Patrick Grim
Stuart Hampshire
Yuval Harari
William Harvey
Kaspar Hauser
Werner Heisenberg
Hodges
Ted Honderich
David Hume
Dale Jaquette
Thomas Jessop
Kant
G. Kasparov
John Keats
G. Leibniz
Leonardo da Vinci
Mona Lisa
Primo Levi
Clarice Lispector
John Locke
Rosa Luxemburg
Henry Marsh
Mathey
J.S. Mill
G.E. Moore
Sidney Morganbesser

W. Mozart
Thomas Nagel
John Nash
Graham Oppy
Anthony Palmer
Samuel Palmer
David Papineau
Derek Parfitt
Jackson Pollock
Thomas de Quincey
Wilhelm Reich
J.J. Rousseau
David Runciman
Bertram Russell
Gilbert Ryle
Ahmed Fãris al-Shidy
John Searle
Aaron Sloman
Timothy Sprigge
Nicholas de Stael
Lawrence Sterne
Tom Stoppard
Galen Strawson
P.F. Strawson
Alan Turing
Bernard Williams
Richard Wollheim
Adrian Woolfson
Wolfram

www.ingramcontent.com/pod-product-compliance
Lightning Source LLC
Chambersburg PA
CBHW071302220526
45468CB00001B/233

9781794473256